Activit Metabolomics
and
Mass Spectrometry

Activity Metabolomics

and

Mass Spectrometry

First Edition

Gary Siuzdak

Scripps Research Institute

Cite:
Siuzdak, G.,
Activity Metabolomics and Mass Spectrometry;
MCC Press: San Diego, USA, 2024.
https://doi.org/10.63025/LCUW3037

Copyright © 2024 by MCC Press

All rights reserved

Preface

My first book "Mass Spectrometry for Biotechnology" introduced the capabilities of mass spectrometry and improvements during the period between 1985 and 1994. These developments continued and were largely covered in the 2003 and 2006 editions of The Expanding Role of Mass Spectrometry in Biotechnology. This new book "Activity Metabolomics and Mass Spectrometry" covers some of the same topics providing an overview of mass spectrometry with the inclusion of metabolomics. This reflects our interest in developing mass spectrometry technologies and the corresponding informatics (XCMS and METLIN) for identifying endogenous metabolites that can modulate phenotype and treat disease. This concept of discovering endogenous metabolites to improve health via safe and cost-effective endogenous metabolites is simply building on a concept that has been around for centuries, examples of endogenous metabolites that act as therapeutics include hydrocortisone, progesterone, vasopressin, melatonin, to name a few.

Acknowledgments

I extend my heartfelt thanks to my peers at the Scripps Center for Metabolomics and Mass Spectrometry, whose diligent reviews and edits were invaluable. Special recognition goes to Mirna Tobea who worked on many of the details in putting this book together, and Elizabeth Billings, Winnie (Heim) Uritboonthai, Linh Hoang, Bill Webb, Corey Hoang, and Aries Aisporna for their exceptional dedication. I am deeply grateful for the pioneering work of Martin Giera, Carlos Guijas, Caroline H. Johnson, Oscar Yanes, Julijana Ivanisevic, Gary J. Patti, Ralf Tautenhahn, Colin A. Smith, Richard A. Lerner, Benjamin F. Cravatt, Xavier Domingo-Almenara, and Markus M. Rinschen, whose innovative contributions were pivotal to the success of activity metabolomics. Above all, my profound appreciation is for Mary E. Spilker, my life partner, whose boundless curiosity, and support have made every step of this journey a joy.

Table of Contents

ACTIVITY METABOLOMICS AND MASS SPECTROMETRY .... *i*
PREFACE .... *v*
ACKNOWLEDGMENTS .... *v*
TABLE OF CONTENTS .... *vi*
CHAPTER 1: A HISTORY OF MASS SPECTROMETRY AND METABOLOMICS 1
  PERSPECTIVE 1
  EARLY MASS SPECTROMETRY – THE PHYSICAL ROOTS 2
    Ionization Advances allow for Biological Applications 5
    Small Molecule and Protein Analysis 6
    Metabolomics, Metabolism, & Metabolites 8
      1700s to 1900: Beginning of Metabolite Characterization 9
      1900s: Analytical Technology Drives Discovery 10
  21ST CENTURY OF KNOWNS: METABOLITE IDENTIFICATION 13
    The Road Less Traveled 14
  21ST CENTURY: THE ROAD AHEAD 15
  Overall 17
  HISTORICAL DEVELOPMENTS IN MASS SPECTROMETRY 18
  HISTORICAL DEVELOPMENTS IN METABOLISM/METABOLOMICS ... 19
  USEFUL REFERENCES 20

CHAPTER 2: IONIZATION AND THE MASS SPECTROMETER 25
  PERSPECTIVE 25
  MASS SPECTROMETRY 25
  WHAT IS MASS SPECTROMETRY? 26
    Some Basics 27
    Sample Introduction Techniques 27
  IONIZATION 29
    Protonation 30
    Deprotonation 30
    Cationization 31
    Transfer of a charged molecule to the gas phase 31
    Electron ejection 32
    Electron capture 32
  IONIZATION SOURCES 34
    Electrospray Ionization 35
    Electrospray Solvents 40
    Configuration of the Electrospray Ion Source 41
    Nanoelectrospray Ionization (NanoESI) 41
    Atmospheric Pressure Chemical Ionization 43

  Atmospheric Pressure Photoionization *43*
  Matrix-Assisted Laser Desorption/Ionization (MALDI) *45*
  Desorption/Ionization on Silicon (DIOS) *50*
  Fast Atom/Ion Bombardment *51*
  Electron Ionization *52*
  Chemical Ionization *54*
 NANOSIMS 56
 SUMMARY 58
 QUESTIONS 58
 USEFUL REFERENCES 59

CHAPTER 3: MASS ANALYZERS 60

 PERSPECTIVE 60
 MASS ANALYSIS 60
  Performance Characteristics *60*
  Accuracy *62*
  Resolution (Resolving Power) *62*
  Mass Range *63*
  Tandem Mass Analysis (MS/MS or MSn) *64*
  Scan Speed *64*
 MASS ANALYZERS 65
  Quadrupoles *65*
  Quadrupole Ion Trap *67*
  Linear Ion Trap *68*
  Double-Focusing Magnetic Sector *70*
  Quadrupole Time-of-Flight Tandem MS *71*
  The MALDI with Time-of-Flight Analysis *73*
  Quadrupole Time-of-Flight MS *74*
  Fourier Transform Mass Spectrometry (FTMS) *75*
  Ion Detectors *83*
  Electron Multiplier *83*
  Faraday Cup *84*
  Photomultiplier Conversion Dynode *84*
  Array Detector *85*
  Charge (or Inductive) Detector *85*
 VACUUM IN THE MASS SPECTROMETER 87
 OVERVIEW 88
 QUESTIONS 89
 USEFUL REFERENCES 90

CHAPTER 4: PRACTICAL ASPECTS OF BIOMOLECULAR ANALYSIS 91

 THEN AND NOW 91
 QUANTITATION 92
 CALCULATING MOLECULAR WEIGHT 95
 ISOTOPE PATTERNS 97

SOLUBILITY 98
TIMING 99
CALIBRATION/ACCURACY 99
SAMPLE PURITY AND CLEAN CONTAINERS 102
SENSITIVITY /SATURATION 106
IONIZATION CHARACTERISTICS 107
MATRIX SELECTION/PREPARATION 108
OVERVIEW 110
COMMON QUESTIONS 110

CHAPTER 5: PEPTIDE AND PROTEIN ANALYSIS 113

PERSPECTIVE 113
OVERVIEW OF PEPTIDE AND PROTEIN ANALYSIS 113
PEPTIDES AND PROTEINS BY MALDI 115
   MALDI Sample Preparation Procedures *117*
      Whole proteins 117
      On-Plate Sample Wash 117
   ZipTip™ Sample Wash *118*
PEPTIDES AND PROTEINS BY ELECTROSPRAY 121
   Multiple Charging *123*
   LC/MS *126*
   Declustering *127*
PEPTIDE MASS MAPPING 129
   Identification Using Tandem Mass Spectrometry *133*
SEPARATION 136
   The Requirement for Sample Separation *136*
   Gel Electrophoresis *137*
   2-D Gels *138*
   A Typical in-Gel Digestion Protocol for coomassie stained gels *141*
PROTEIN ID 141
   MALDI-MS *141*
   LC MS/MS *143*
   2D LC-MS/MS *144*
   Protein Mass Mapping and Isotope Labeling *146*
   Protein Profiling *147*
   LC-MALDI MS/MS *148*
OVERVIEW 149
QUESTIONS 156
USEFUL REFERENCES 157

CHAPTER 6: METABOLOMIC FUNDAMENTALS 160

PERSPECTIVE 160
WHAT IS METABOLOMICS AND THE METABOLOME? 160
TARGETED METABOLOMICS 162
UNTARGETED (OR GLOBAL) METABOLOMICS 163
METABOLITE CHARACTERIZATION 167
CHALLENGES 168

OVERVIEW	172
USEFUL REFERENCES	172

CHAPTER 7: XCMS DATA PROCESSING 173

PERSPECTIVE	173
HOW ARE METABOLOMICS DATA PROCESSING PLATFORMS USED?	174
XCMS DATA STREAMING	175
XCMS DATA ANALYSIS EXAMPLE	177
XCMS CLOUD PLOT	178
XCMS META ANALYSIS	179
WHAT'S GOING ON INSIDE XCMS	180
XCMS WORKFLOW	181
PEAK PICKING	*183*
MATCHED FILTER PEAK PICKING	*183*
CENTWAVE PEAK PICKING	*184*
PEAK ALIGNMENT	*185*
FILL PEAK	*187*
XCMS R PACKAGE	188
XCMS GUIDED SYSTEMS BIOLOGY	189
XCMS MRM	190
OVERVIEW	191
REFERENCES	191

CHAPTER 8: METLIN 192

PERSPECTIVE	192
METLIN-MS/MS	193
METLIN-MS/MS Basic Search Engines	196
METLIN Identification via MS/MS Matching	198
METLIN - NL	201
METLIN - Iso	203
METLIN - RT	204
METLIN - MRM	204
METLIN - CCS	207
OVERVIEW	208
REFERENCES	209

CHAPTER 9: METABOLITE DISCOVERY 210

PERSPECTIVE	210
ELEMENTAL COMPOSITION STRUCTURE DETERMINATION	211
IDENTIFYING METABOLITES/LIPIDS (OLEAMIDE): SLEEP	215
Identifying Lipid Neuroprotectin D1: Stem Cells and the "Plastic Metabolome"	*216*
Identifying Metabolite 3-Indole Propionic Acid: Gut Microbiota	*218*
Identifying Lipid N,N-Dimethylsphingosine: Chronic Pain	*219*
IDENTIFYING METABOLITE N1,N12-DIACETYLSPERMINE: BACTERIA BIOFILMS CANCER	221

viii

OVERVIEW	223
REFERENCES	224

CHAPTER 10: ACTIVITY METABOLOMICS 225

PERSPECTIVE	225
HOW DO "ACTIVE METABOLITES" WORK?	227
METABOLIC MODIFICATION OF DNA, RNA AND PROTEINS	228
METABOLITE-MACROMOLECULE INTERACTIONS	229
DISCOVERING ACTIVE METABOLITES	231
METABOLIC ACTIVITY SCREENING STRATEGIES	232
MULTI-OMICS INTEGRATION FOR DETERMINING ACTIVITY	235
PERSPECTIVE: APPLICATIONS OF ACTIVITY METABOLOMICS	237
CONCLUSIONS	239
Endogenous Metabolite Success Stories	240
REFERENCES	241

INDEX ....

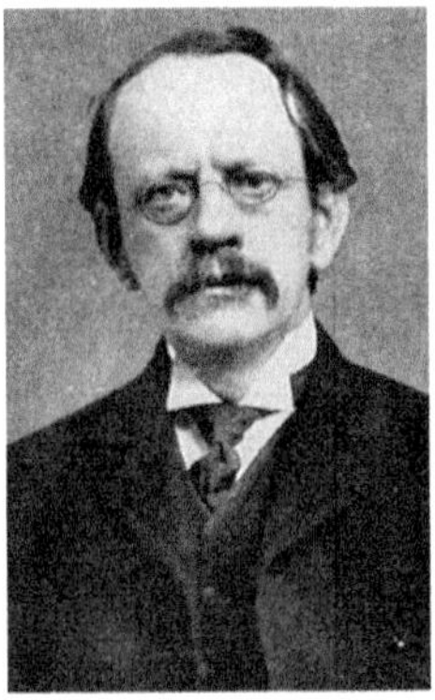
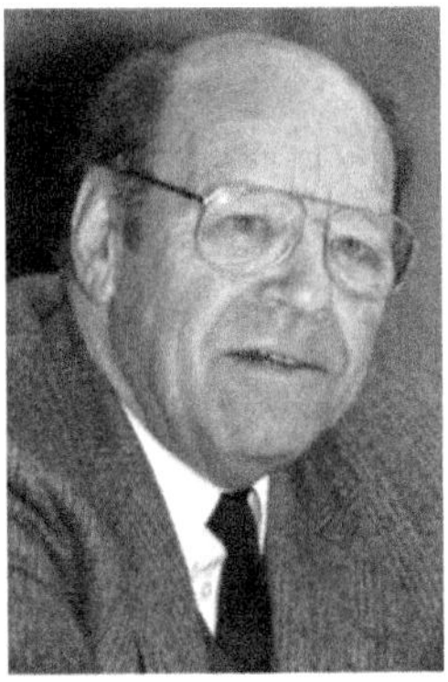
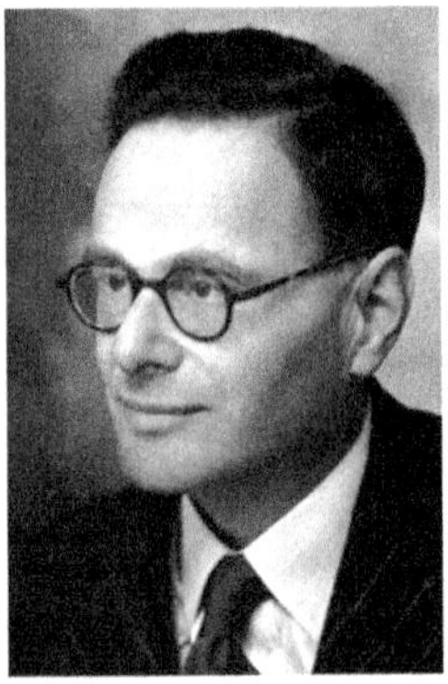

Chapter 1

A History of Mass Spectrometry and Metabolomics

Mass Spectrometry

Mass spectrometry has a special intrigue that comes from its interdisciplinary nature, as it freely crosses the borders of physics, chemistry, and biology. Its history goes back to the early 1900's, and its development has reached a pinnacle in recent years, being widely applied to a range of problems. Thousands of women and men have contributed to the development of mass spectrometry, and this timeline highlights some key individuals and events. Metabolomics and the study of metabolism has also had an equally interesting trajectory encompassing many different technologies, however it is safe to say that mass spectrometry has been a key technology in its evolution (**Figure 1.1**).

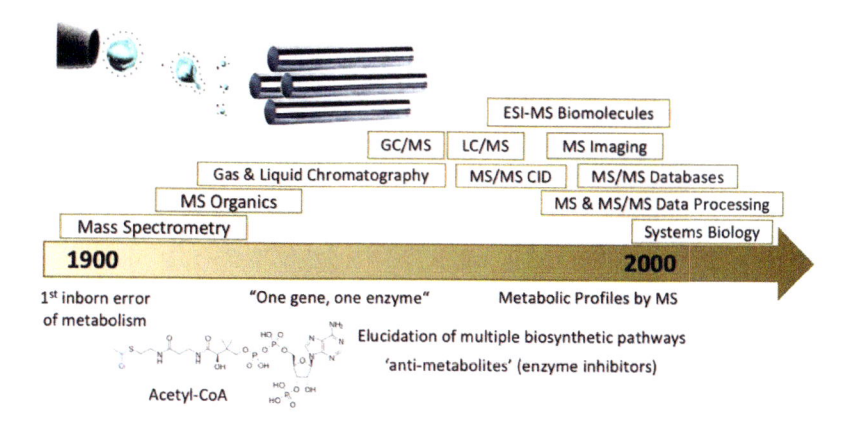

Figure 1.1. A broad view of how technological developments coincided with biochemical discoveries and the development of metabolomics.

Early Mass Spectrometry – The Physical Roots

The history of MS begins with J.J. Thomson (**Figure 1.2**) of the University of Cambridge. Thomson's "theoretical and experimental investigations on the conduction of electricity by gases" led to the discovery of the electron in 1897 for which he was awarded the 1906 Nobel Prize in Physics. In the first decade of the 20th century Thomson went on to construct the first mass spectrometer (then called a parabola spectrograph), in which ions were separated by their different parabolic trajectories in electromagnetic fields and detection occurred when the ions struck a fluorescent screen or photographic plate.

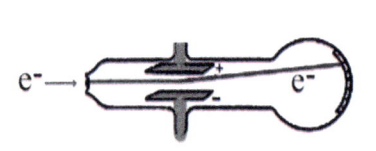

Figure 1.2. Thomson's cathode ray tube

J.J. Thomson
1906 Nobel Prize in Physics

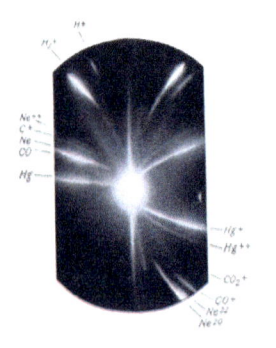

Figure 1.3. Aston's photograph of ions deflected by electric and magnetic field

Francis William Aston
1922 Nobel Prize in Chemistry

Later, Thomson's protégé, Francis W. Aston (1922 Nobel Prize in Chemistry, **Figure 1.3**) of the University of Cambridge, designed a mass spectrometer that improved resolving power, allowing Aston to study isotopes. During this same period, A.J. Dempster of the University of Chicago also improved upon resolution with a magnetic analyzer and developed the first electron ionization source, which ionized volatilized molecules with a beam of electrons. Electron ionization ion sources are still very widely used in modern mass spectrometers for small molecule analysis. Thomson, Aston, and Dempster created a strong foundation of MS theory and instrument design that made it possible for those who followed to develop instruments capable of meeting the demands of chemists and biologists.

An important goal, particularly for chemists, was to create an instrument capable of analyzing both the elements and small organic molecules. Eventually this led to mass analyzers of five different forms: magnetic sector double-focusing, time-of-flight, quadrupole, Fourier transform ion cyclotron resonance and fourier transform orbitrap mass analyzers. In its time, the most widely used high mass-resolution double-focusing instrument was developed by Professor Alfred O.C. Nier at the University of Minnesota during World War II to perform isotopic analysis and separate uranium-235 from uranium-238. The first nuclear bomb was developed entirely from the uranium separated by this type of mass spectrometer.

The concept of time-of-flight mass spectrometry (TOF MS) was proposed in 1946 by William E. Stephens of the University of Pennsylvania. In a TOF analyzer, ions are separated by differences in their velocities as they move in a straight path toward a collector. TOF MS allows for rapid analyses (5-20 kHz repetition rates), it is capable of high resolving power and high accuracy, it is applicable to rapid chromatographic detection, and it is now used for the mass determination of large biomolecules because of its virtually limitless mass range.

One type of mass analyzer that proved to be ideal for coupling to gas chromatography and liquid chromatography (in the 1990s) was the quadrupole mass filter. The quadrupole was first reported in the mid-1950s by Wolfgang Paul (**Figure 1.4**), who shared the 1989 Nobel Prize in Physics for his work on ion trapping. In a quadrupole device, a quadrupolar electrical field (comprising radiofrequency and direct-current components) is used to separate ions. Although quadrupole mass analyzers are not as accurate and precise as double-focusing instruments, they offer excellent dynamic range, are quite stable, and are also readily applied to tandem mass spectrometry experiments – features that make them popular for quantitative analysis and drug discovery applications. Quadrupoles are the most widely used mass analyzers.

Figure 1.4. Paul's quadrupole and quadrupole ion trap resulted in two of the most used mass analyzers.

Wolfgang Paul
1989 Nobel Prize in Physics

Ion cyclotron resonance MS (ICR-MS) and the orbitrap have proven to be a powerful MS solutions for high resolution and high accuracy. ICR was initially described by John A. Hipple of the National Bureau of Standards, Washington, D.C. It operates by subjecting ions to a simultaneous radiofrequency electric field and a uniform magnetic field, causing the ions to follow spiral paths in an analyzer chamber. By scanning the radiofrequency or magnetic field, the ions can be detected sequentially. In 1974, Melvin B. Comisarow and Alan G. Marshall of the University of British Columbia revolutionized ICR by developing Fourier transform ICR mass spectrometry (FTMS). The major advantage of FTMS is that it allows many different ions to be measured at once. Sub-part per million accuracy – such as an error of less than 0.001 Dalton for a 1000 Dalton peptide – is now routinely possible with commercial FTMS instruments. The orbitrap, released by Makarov in 2005, was the first high resolution and sensitivity trapping type of mass analyzer that employed electrostatic fields which also employs FT data deconvolution.

Ionization Advances allow for Biological Applications

A variety of mass analyzer and tandem mass spectrometry designs are widely used today and are continually being developed for new applications. Despite advances in mass accuracy, mass range, quantitative analysis, and the ability to couple the instruments to chromatography, MS still lacked efficacy for large and small biomolecule analysis. Significant molecular decomposition or fragmentation during vaporization/ionization and poor sensitivity proved very problematic in biomolecular MS. The development of "soft ionization" – electrospray ionization MS (ESI MS) and matrix-assisted laser desorption/ionization MS (MALDI MS) – allowed MS to transcend into the realm of biology (**Figure 1.5**).

John Fenn - 2002 Nobel Prize

Koichi Tanaka -2002 Nobel Prize

Figure 1.5. ESI and soft laser desorption/ionization MS revolutionized the nondestructive analysis of biomolecules, ushering in MS being widely applied in the biological sciences.

In ESI MS, highly charged droplets dispersed from a capillary in an electric field are evaporated, and the resulting ions are drawn into an MS inlet. The technique was first conceived in the 1960s by chemistry professor Malcolm Dole of Northwestern University but was not put into practice for biomolecule analysis until the early 1980s by John B. Fenn of Yale University.

In soft laser desorpton/ionization and matrix-assisted laser desorption/ionization (MALDI) MS, analyte molecules are laser-desorbed from a solid or liquid UV-absorbing matrix. The technique was initially reported by Koichi Tanaka and coworkers for proteins at Shimadzu Corp., Kyoto, Japan, and was also developed by Franz Hillenkamp and Michael Karas at the University of Frankfurt, Germany.

For their work on developing soft ionization techniques suitable for large biomolecule analysis, Fenn and Tanaka shared the 2002 Nobel Prize in Chemistry. Their work on ESI-MS and MALDI-MS made MS increasingly useful for sophisticated biological experiments. Applications include protein identification, drug discovery, DNA sequencing, carbohydrate analysis, and biomarker discovery.

Small Molecule and Protein Analysis

The importance of these MS developments can be seen in research efforts at every major pharmaceutical company and university. Arguably the most important MS application is in small molecule and peptide quantification using triple quadrupoles (QqQ) MS (**Figure 1.6**), the most widely used mass spectrometer. Endogenous biomolecules, exogenous molecules and peptides are routinely quantified with this technology. Perhaps most compelling is that virtually all newborns in the United States as well as elsewhere are tested with QqQs for over 30 different diseases.

ESI QqQ triple quadrupole multiple reaction monitoring

ESI Q single quadrupole multiple reaction monitoring (using enhanced in-source fragmentation)

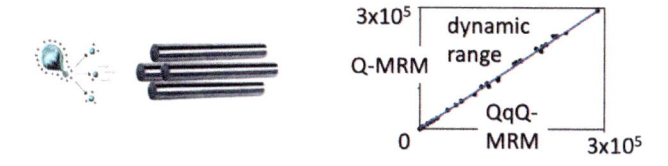

Figure 1.6. Triple and single quadrupole) multiple reaction monitoring for small molecule and peptide quantification.

At the Limits of Mass Spectrometry: Viruses

Intact Viral Mass Analysis

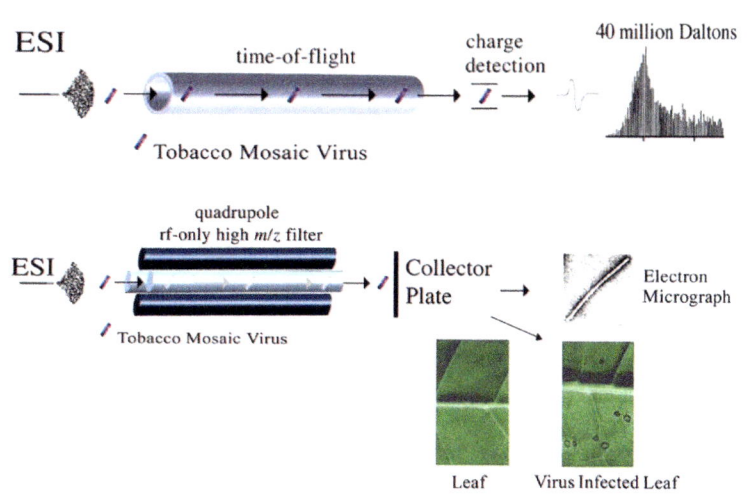

Figure 1.7. (top) Intact-virus analysis with a TOF charge-detection MS. (bottom) ESI virus selection, capture, and viability experiments. The native virus structure and viability was preserved. *Chemistry & Biology 1996 & Angew. Chemie 2001.*

   Soft ionization ESI MS methods have made it possible to study increasingly larger structures and their noncovalent interactions. ESI instrumentation designed by Henry Benner at Lawrence Berkeley Laboratory was the first to measure intact viral ions measuring millions of Daltons, and earlier experiments show that viruses maintain their structure and virulence (**Figure 1.7**).

Metabolomics, Metabolism, & Metabolites

The 100+ year history of mass spectrometry pales in comparison to the centuries long history of metabolism. The discovery of metabolites and the subsequent field of metabolomics has played a critical role in advancing biochemistry. Hermann Boerhaave, an 18th-century scientist, and author of *Elementa Chemiae* characterized the first metabolite, urea, which set the stage for identifying the other most abundant metabolites such as amino acids, sugars, and lipids over the next two centuries. However, with the emergence of advanced analytical technologies from the 1930s onwards, finer details of metabolism began to surface. These technologies provided a much clearer and detailed image of biochemistry, uncovering its structural complexity and revealing new metabolic activities.

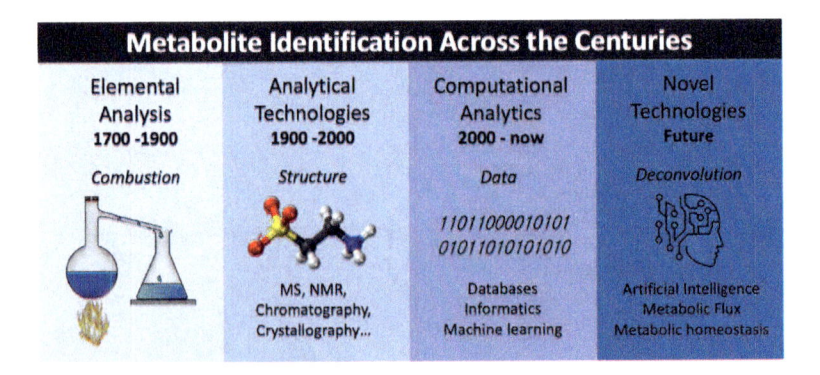

Figure 1.8. The evolution of metabolic sciences over the centuries.

Recent advances (**Figure 1.8**) have revitalized the importance of metabolism in many biological and therapeutic fields, including cancer, bioengineering, systems biology, the microbiome, and nutrition, among others. The integration of metabolomics, that includes a suite of technologies that characterize and quantify metabolites contributing to the structure, function, and dynamics of an organism, has provided an additional tool for classical biochemistry, enabling further advances in biochemical research. Mass spectrometry technology, particularly the Nobel Prize-winning ESI, has made it possible to routinely observe intact molecular ions, a capability previously unachievable for most biomolecules. In addition, ESI has enabled sensitivity several orders of magnitude greater. These breakthroughs have facilitated the discovery of metabolites that were not previously known to exist in nature, including those found in multicellular organisms such as animals and plants, as well as unicellular microorganisms such as bacteria.

Determining the structures of unknown metabolites is a significant technological challenge, but it also presents an exciting opportunity to expand our biochemical knowledge base for both characterized and uncharacterized organisms. The identification of all known and unknown small molecules in an organism, referred to as the metabolome, represents the future of metabolism discoveries. From the 1700s to the 1900s, newly discovered metabolites were identified due to their ubiquitous nature. Their abundance made them easy enough to be isolated and characterized by the analytical techniques available at that time (for example amino acids, monosaccharides, and nucleotides). However, in recent decades, the discovery of new and less ubiquitous metabolites and classes of metabolites has opened new avenues of research and led to a sustained effort to characterize and understand their physiological effects.

1700s to 1900: Beginning of Metabolite Characterization

Take some very fresh well-concocted Urine of persons in perfect Health, put it preferentially into a very clean Vessel, and with an equable Heat of 200 degrees, evaporate it till you have reduced it to the consistence of fresh Cream" ..."Put a large quantity of this thick inspissated Liquor into a tall cylindrical glass vessel with a paper tied over it and let it stand quite in a cool place for the space of a year..." – Hermann Boerhaave (Duranton et al., 2016)

The analysis of small organic molecules, such as lactic acid, urea, oxalic acid, and citric acid, dates back to the 1700s when analytical techniques developed by Boerhaave and Lavoisier were applied. These methods were later improved by Gay-Lussac and Thenard in the early 1800s. The analysis typically involved the isolation and purification of specific molecules from animal and food products, such as citric acid from lemons and lactic acid from fermented milk, using techniques such as distillation and crystallization. The atomic weights of the constituents were then derived through combustion analysis. While these chemical formulas were informative, they only allowed for structural hypotheses. Nonetheless, the early work on these molecules laid the foundation for subsequent advances in analytical chemistry and paved the way for modern metabolomics technologies.

During the 19th century, significant progress was made in establishing the molecular formulas of many metabolites (Thaulow, 1838). However, one of the most significant breakthroughs of this era was the publication of Justus von Liebig's book "Animal Chemistry (Die Thier-Chemie)" (Freiherr von Liebig, 1843). Liebig's research laid the foundation for our knowledge of metabolic reactions and the inter-conversions of simple organic molecules within cells. In the book, Liebig inferred metabolic equations that described physiological processes based solely on his knowledge of organic chemistry, without any evidence of their existence in vivo. This groundbreaking work allowed for the analysis of metabolic pathways and their regulation, paving the way for further advances in biochemical research.

1900s: Analytical Technology Drives Discovery

The 1900s ushered in a remarkable era of technological advancements that played a pivotal role in understanding the central carbon metabolism (Krebs cycle, **Figure 1.9**) of eukaryotic cells. The utilization of nuclear reactors as a source of artificial radioisotopes and the development of x-ray crystallography and scintillation spectrometers greatly accelerated biochemical

1. A History of Mass Spectrometry and Metabolomics

research in the 1930s and beyond. The development of nuclear magnetic resonance (NMR) and mass spectrometry (MS) greatly advanced biochemical methods and metabolomics principles. NMR was first described by Isidor Rabi in 1938 and later used for the analysis of liquids and solids by Bloch and Purcell. Simultaneously, chromatographic techniques developed by Archer Martin and Richard Synge, such as gas chromatography and high-pressure liquid chromatography (HPLC), furthered the pace of research in the field.

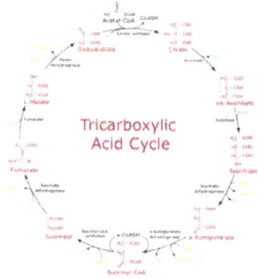

Figure 1.9. Krebs cycle, aka the tricarboxylic acid (TCA) cycle.

Hans Krebs 1953 Nobel Prize

By 1945, most of the analytical techniques necessary for biochemical research were already available to the next generation of researchers. As a result, by 1957, biosynthetic pathways for virtually all types of known biological molecules had been elucidated, including lipids, carbohydrates, nucleic acid bases, amino acids, and vitamins. In fact, in 1955, Donald Nicholson compiled all of this knowledge into a single map comprising approximately 20 metabolic pathways. Thus, molecular formulas and metabolite structures had already been discovered well before the first structure of a protein (myoglobin) with atomic resolution, the elucidation of the DNA structure in 1953, or the subsequent publication in 1958 of molecular biology's central dogma. However, the most important advances in mass spectrometry-based metabolomics were made by Gohlke et al., McLafferty et al. who introduced collision induced dissociation (CID) in 1968, and the coupling of liquid chromatography to mass spectrometry in 1974.

In the late 1940s and early 1950s, Roger J. Williams introduced the concept of individual biochemical profiles. The first successful demonstration of mass spectrometry-based metabolomics was reported in 1966 by Dalgliesh and colleagues, who used GC/MS to separate and detect a wide range of metabolites in urine and biological tissue extracts. The term "metabolic profiles" was later introduced by Horning and colleagues, who, together with Linus Pauling and Arthur Robinson, developed GC/MS methods to simultaneously monitor dozens of metabolites in biological samples during the 1970s. This work was followed by important contributions from Gates and Sweeley, who established GC/MS as a quantitative tool for metabolic profiling.

Analytics, Databases & Bioinformatics in Identification

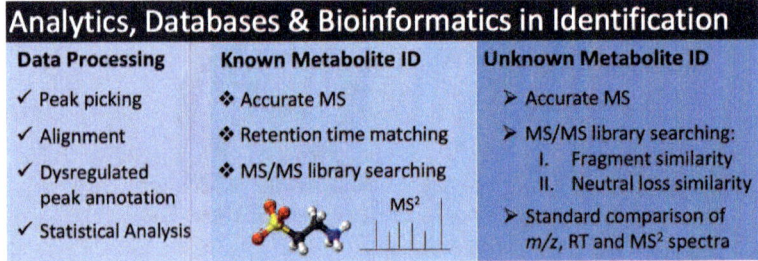

Data Processing	Known Metabolite ID	Unknown Metabolite ID
✓ Peak picking	❖ Accurate MS	➢ Accurate MS
✓ Alignment	❖ Retention time matching	➢ MS/MS library searching:
✓ Dysregulated peak annotation	❖ MS/MS library searching	I. Fragment similarity
✓ Statistical Analysis		II. Neutral loss similarity
		➢ Standard comparison of m/z, RT and MS^2 spectra

Figure 1.10. The analytics and data processing involved in metabolite ID including peak detection, alignment, deconvolution, and spectral matching via MS/MS databases of standards, and statistical assessment of metabolomic data.

The development of electrospray ionization (ESI) by John B. Fenn in 1989 marked a significant milestone in the field of metabolomics (and proteomics). The advent of ESI paved the way for the first LC-ESI MS-based untargeted metabolomics studies in 1994, leading to the discovery of the vast potential of untargeted metabolite profiling. However, these initial experiments also highlighted several mass spectrometric improvements that needed to be addressed to enable effective interpretation of untargeted mass spectral data and the identification of unknown metabolites. Specifically, there was a need for peak detection and alignment in convoluted LC/MS spectra to statistically characterize meaningful metabolic features, as well as the need for tandem mass spectrometry databases to facilitate rapid identification of unknown metabolites (**Figure 1.10**).

21st Century Knowns: Metabolite Identification

In the 1800s and 1900s, the identification of metabolites was carried out one at a time by isolating and characterizing significant amounts of a single target compound from natural sources. In contrast, modern 'omic' technologies and metabolomics are comprehensive by design, aiming to characterize and quantify all metabolites collectively. However, the mass spectral data generated from these technologies produces a highly complex dataset on thousands of molecules, each with adducts, isotopes, in-source fragments, background, and contaminant ions. Consequently, there is a distinction between metabolite annotation and identification. Annotation refers to the assignment of a candidate metabolite to multiple and redundant MS signals based on analytical characteristics such as retention time and m/z, whereas identification requires a much more tedious and conclusive process, assigning a chemical structure to a candidate metabolite through comparison with chemically pure standard materials or conclusive (2D) NMR data.

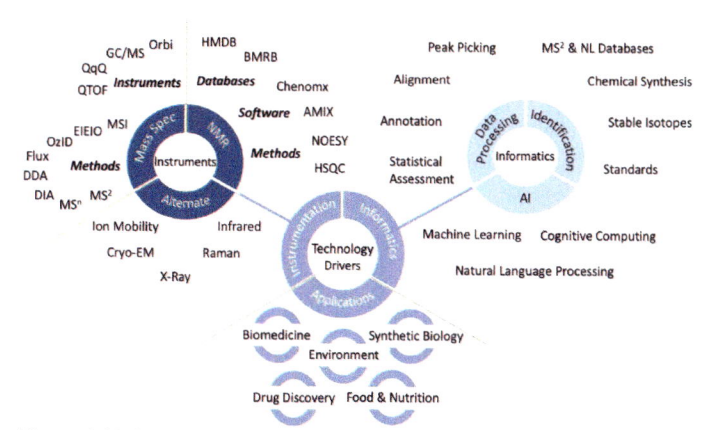

Figure 1.11. It's complicated. A compendium of technologies that illustrate the state-of-the-art in deciphering metabolomics data.

As a result (**Figure 1.11**), the 21st century has seen the development of increasingly sophisticated signal processing techniques for MS and NMR-based metabolomics, enabling the annotation and identification of vast amounts of highly complex data. This includes a multitude of peak detection and alignment software. Tandem (MS/MS) mass spectral data are used for metabolite identification, which can produce structural information for hundreds or thousands of metabolites in minutes. MS/MS methods are continually evolving and improving to match experimental data with spectral databases.

Spectral databases play a critical role in the metabolite annotation process. For decades, GC/MS has been the dominant identification technology due to the vast size of its chemical mass spectral libraries. For instance, the National Institute for Standards and Technology (NIST) has an impressive library of electron ionization (EI) mass spectra generated from over 300,000 compounds. However, since the advent of EI, MS technologies have undergone significant advancements. This progress was recognized by the 2002 Nobel Prizes, which acknowledged the development of two ionization approaches enabling the detection of intact biomolecules (ESI and soft laser desorption ionization). ESI has since become the dominant technology due to its "softer" nature, making it less destructive and compatible with LC separation methods. Nevertheless, the availability of publicly available MS/MS spectra for small molecules and metabolites was initially limited for LC-ESI MS approaches. The creation of the first database of ESI MS/MS spectra specifically designed for the identification of small molecules and metabolites in 2003 addressed this gap. Since then, numerous public and commercial databases and spectral libraries have been established, including METLIN, which now has MS/MS experimental data on over 900,000 molecular standards.

21st Century of Unknowns: The Road Less Traveled

Metabolomics scientists often face the daunting task of characterizing unknown unknowns, which involves identifying metabolites with unknown structures and functional roles. Major technological advancements are required to answer the many intriguing questions surrounding metabolism research. Assigning identities to the thousands of spectral signals from metabolome-wide studies is a time-consuming process, and the characterization of unknown metabolites presents a significant bottleneck. Accurately identifying metabolites is crucial as researchers cannot afford to allocate precious resources to research on wrongly assigned metabolic structures. To fully comprehend the metabolome, it is essential to determine the origin, fate, and functional roles of each metabolite, which requires significant technological advancements. Therefore, prioritizing the discovery and identification of metabolites is imperative to gain a comprehensive understanding of metabolism. Although, on a bright note, it is coming to light that many of these signals are in- source fragments, thus, the picture may not be as complicated as we once thought.

21st Century: The Road Ahead

At the turn of the 21st century, advances in technology and bioinformatics enabled the large-scale investigation of genomic, transcriptomic, and proteomic data, leading to the characterization of DNA/RNA and proteins. However, the study of metabolites proved to be more challenging as they are not comprised of monomers and required different approaches. It was not until the last two decades that advancements in LC-MS/MS and bioinformatics allowed for comprehensive, metabolome-wide investigations, leading to the discovery of new metabolites and an increased coverage of the metabolome, despite the challenges posed by the dynamic range of endogenous metabolites (**Figure 1.12**). A bottleneck in metabolomics has been the availability of a comprehensive database of tandem mass spectra for metabolites, but this issue is currently being addressed by multiple initiatives including METLIN.

21st Century: The Road Ahead

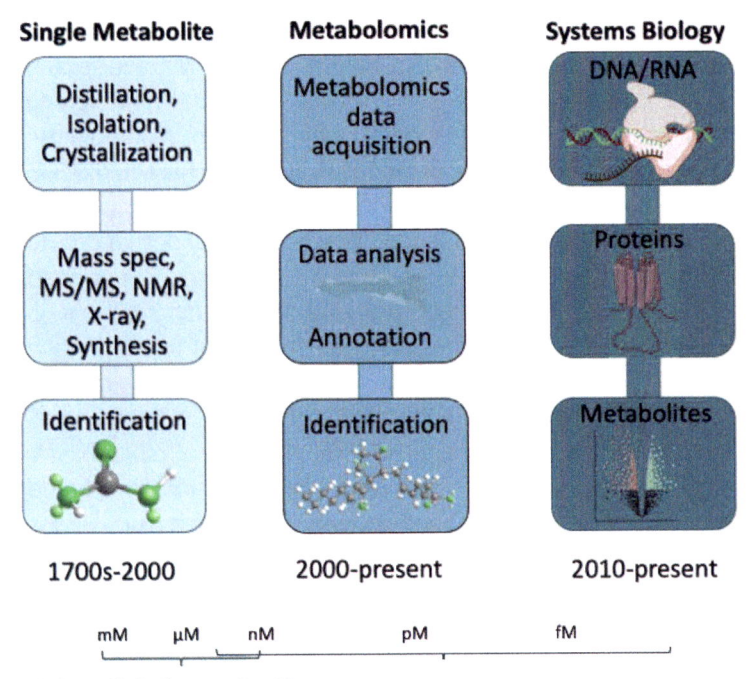

Figure 1.12. An evolution in metabolism and metabolite identification.

Natural language processing, NLP, a branch of AI (**Figure 1.13**), will likely help annotate molecules and identify biochemical relationships between them even when there are no reference spectra available. By embedding metabolic features in their biological context, we can improve the accuracy of annotations and shorten the list of possibilities. This can be achieved by cognitive literature mining, which uses NLP and machine learning to extract key concepts from scientific literature and predict potential connections between entities.

Organismal Level Deconvolution

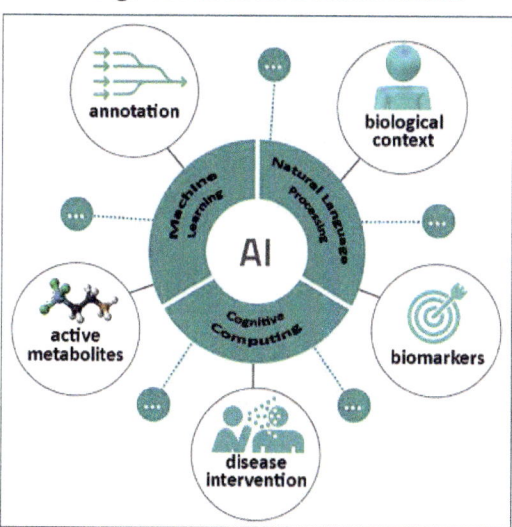

future

Figure 1.13. Cognitive computing and AI are altering the way big data are processed, providing immediate literature-based contextualization during the prioritization, identification, and interpretation steps in the metabolomics data analysis.

Overall

Over the last two centuries, the fields of mass spectrometry and metabolism have experienced a remarkable co- evolution. Starting from early efforts to understand the composition of gases and molecules, mass spectrometry has advanced tremendously and become a cornerstone of modern analytical chemistry. At the same time, the study of metabolism has undergone a similar revolution, with the development of new techniques and tools that allow for a more detailed understanding of the complex biochemical processes that occur within living organisms. Together, these two fields have enabled a deeper understanding of the fundamental principles that govern life, and have paved the way for new breakthroughs in medicine, biology, and many other fields. As we continue to push the limits of what is possible with mass spectrometry and metabolism, it is clear that these two disciplines will continue to shape our understanding of the world around us for many years to come.

Urea was the first piece of the puzzle in understanding metabolism, and centuries later, new discoveries are still being made thanks to technological advances especially in mass spectrometry. It is "just" a matter of characterizing all the endogenous and exogenous molecules, a daunting yet exciting effort that will continue for decades to come.

Historical Developments in Mass Spectrometry

Investigator(s)	Year	Contribution
J.J. Thomson	1899-1911	First Mass Spectrometer
Dempster	1918	Electron Ionization and Magnetic Focusing
Aston	1919	Atomic Weights using MS
Mattauch & Herzog	1934	Double Focusing Instruments
Stephens	1946	Time-of-Flight Mass Analysis
Hipple, Sommer & Thomas	1949	Ion Cyclotron Resonance
Johnson & Nier	1953	Reverse Geometry Double Focusing Instruments
Paul & Steinwedel	1953	Quadrupole Analyzers
Beynon	1956	High Resolution MS
McLafferty and Ryahe	1959-1963	GC/MS
Biemann, Cone, Webster, & Arsenault	1966	Peptide sequencing
Munson & Field	1966	Chemical Ionization
Dole	1968	Electrospray Ionization

1. A History of Mass Spectrometry and Metabolomics

Investigator(s)	Year	Contribution
Beckey	1969	Field Desorption-MS of Organic Molecules
MacFarlane & Torgerson	1974	Plasma Desorption-MS
Comisarow & Marshall	1974	FT ICR MS
Yost & Enke	1978	Triple Quadrupole MS
Barber	1981	Fast Atom Bombardment (FAB)
Fenn	1984	ESI on Biomolecules
Tanaka, Karas, & Hillenkamp	1985-8	Matrix facilitated laser desorption/ionization
Chowdhury, Katta & Chait	1990	Protein Conformational Changes with ESI-MS
Ganem, Li, & Henion Chait & Katta	1991	Noncovalent Complexes with ESI-MS
Pieles, Zurcher, Schär, & Moser	1993	Oligonucleotide ladder Sequencing
Henzel, Billeci, Stults, Wong, Grimley, & Watanabe	1993	Protein Mass Mapping
Lerner, Siuzdak, Prospero-Garcia, Henriksen, Boger, Cravatt	1994	LC/MS based metabolomics
Makarov	2000	Orbitrap

Historical Developments in Metabolism/Metabolomics

Investigator(s)	Year	Contribution
Greece	300	Body fluids used to predict disease
Santorio Sanctorius	1614	Quantitative basis of metabolism
Thomas Willis	1674	Diabetes measured qualitatively by urine sweetness
Matthew Dobson	1776	Sugar identified in diabetic urine

Investigator(s)	Year	Contribution
J.J. Thomson	Early 1900s	First mass spectrometer
Krebs and Henseleit	1932	Urea cycle
Krebs	1937	Krebs cycle
Woolley and White	1943	"antimetabolite" aka inhibitors
Bloch and Purcell	1946	NMR introduced
Williams et al.,	1951	Metabolic patterns in human health
Allan et al.,	1958	Argininemia
Westall	1960	Argininosuccinic Aciduria
Dalgliesh et al.,	1966	MS-Based Metabolomics (GC/MS)
Mamer, Hornig, Sweeley et al.,	1971	Human Metabolite MS Profiling
Hoult et al.,	1974	NMR based metabolomics
Millington, Chace et al.,	1990	Clinical Tandem MS
Lerner, Siuzdak et al.,	1994	Untargeted LC/MS Metabolomics
Oliver et al.,	1998	"Metabolome" coined
Gross et al.,	2003	Shotgun Lipidomics

Useful Mass Spectrometry References

Thomson JJ. On the Masses of the Ions in Gases at Low Pressures. Philosophical Magazine. 1899, 48:295, 547-567.

Thomson JJ. On Rays of Positive Electricity. The London, Edinburgh, and Dublin Philosophical Magazine and Journal of Science. 1907, XLVII.

Thomson JJ. Rays of Positive Electricity. Phil. Mag. 1911, 6:20, 752-67.

Dempster AJ. A new method of positive ray analysis. Phys. Rev. 1918, 11, 316- 24.

Aston FW. A Positive Ray Spectrograph (Plate IX). London, Edinburgh and Dublin Philosophical Magazine and Journal of Science. 1919, 6:38:228, 709.

1. A History of Mass Spectrometry and Metabolomics

Aston FW. Isotopes and Atomic Weights. Nature. 1920, 105, 617.

Aston FW. The Mass-Spectra of Chemical Elements. Phil. Mag. 1920, 39, 611-25.

Stephens W. Phys. Rev. 1946, 69, 691

Hipple JA, Sommer H, Thomas HA. A Precise Method of Determining the Faraday by Magnetic Resonance. Phys. Rev. 1949, 76, 1877-1878.

Nier AO. A double-focusing mass spectrometer. Natl. Bur. Standards(U.S.) Circ. 1953, 522, 29-36.

Paul W, Steinwedel H. A new mass spectrometer without magnetic field. Z. Naturforsch. 1953, 8A, 448-450.

Dalgliesh, C.E., Horning, E.C., Horning, M.G., Knox, K.L., and Yarger, K. (1966). A gas-liquid-chromatographic procedure for separating a wide range of metabolites occuring in urine or tissue extracts. Biochem J 101, 792-810.

Dole M, Mack LL, Hines RL, Mobley RC, Ferguson LD, Alice MB. Molecular beams of macroions. Journal of Chemical Physics. 1968, 49:5, 2240-2249.

Comisarow MB, Marshall AG. Fourier transform ion cyclotron resonance [FT-ICR] spectroscopy. Chem. Phys. Lett. 1974, 25:2, 282-283.

Tanaka K, Waki H, Ido Y, Akita S, Yoshida Y., Yoshida T. Protein and polymer analysis up to m/z 100,000 by laser ionization time-of-flight mass spectrometry. Rapid Commun. Mass Spectrom. 1988, 2, 151.

Karas M, Hillenkamp F. Laser desorption ionization of proteins with molecular mass exceeding 10,000 Daltons. Anal. Chem. 1988, 60, 2299-2301.

Fenn JB, Mann M, Meng CK, Wong SF, Whitehouse CM. Electrospray Ionization for Mass Spectrometry of Large Biomolecules. Science. 1989, 246, 64-71.

Ganem B, Li YT, Henion JD. Detection of Noncovalent Receptor Ligand Complexes by Mass Spectrometry. Journal of the American Chemical Society. 1991, 113:16, 6294-6296.

Katta V, Chait BT. Conformational Changes In Proteins Probed By Hydrogen-Exchange Electrospray-Ionization Mass Spectrometry. Rapid Communications In Mass Spectrometry. 1991, V5:N4, 214-217.

Katta V, Chait BT. Observation Of The Heme Globin Complex In Native Myoglobin By Electrospray-Ionization Mass Spectrometry. Journal Of The American Chemical Society. 1991, V113:N22, 8534-8535.

Henzel WJ, Billeci TM, Stults JT, Wong SC, Grimley C, Watanabe C. Identifying Proteins From 2-Dimensional Gels By Molecular Mass Searching of Peptide Fragments in Protein Sequence Databases. Proceedings Of The National Academy Of Sciences Of The United States Of America. 1993, V90:N11, 5011- 5015.

Siuzdak G, Bothner B, Yeager M, Brugidou C, Fauquet CM, Hoey K., Chang C.M. Mass Spectrometry and Viral Analysis. Chemistry & Biology. 1996, 3, 45.

Yergey AL, Yergey AK. Preparative Scale Mass Spectrometry: A Brief History of the Calutron. JASMS. 1997, V8:N9, 943-953.

Makarov, A. Electrostatic axially harmonic orbital trapping: A high-performance technique of mass analysis. Analytical Chemistry 2000. 72 (6): 1156–62.

Fuerstenau SD, Benner WH, Thomas JJ, Brugidou C, Bothner B, Siuzdak G. Mass Spectrometry of an Intact Virus. Angewandte Chemie. 2001, 40, 542-544.

Chace, DH. Mass Spectrometry in the Clinical Laboratory. Chem. Rev. 2001, 101, 445–477.

Useful Metabolism/Metabolomic References

Chevreul, M.E. (1823). A chemical study of oils and fats of animal origin. (St Eutrope-de-Born: Sàrl Dijkstra-Tucker).

Rutherford, E., and Geiger, H. (1908). An electrical method of counting the number of α-particles from radio-active substances. Proceedings of the Royal Society of London. Series A, Containing Papers of a Mathematical and Physical Character 81, 141-161.

Windaus, A. (1932). Über die Konstitution des Cholesterins und der Gallensäuren. Hoppe-Seyler's Zeitschrift Fur Physiologische Chemie 213, 147-187.

Chiewitz, O., and Hevesy, G. (1935). Radioactive Indicators in the Study of Phosphorus Metabolism in Rats. Nature 136, 754-755.

Krebs, H.A. (1940). The citric acid cycle and the Szent-Györgyi cycle in pigeon breast muscle. Biochem J 34, 775-779.

Beadle, G.W., and Tatum, E.L. (1941). Genetic Control of Biochemical Reactions in Neurospora. Proceedings of the National Academy of Sciences 27, 499-506.

Woodward, R.B., Sondheimer, F., and Taub, D. (1951). THE TOTAL SYNTHESIS OF CHOLESTEROL. Journal of the American Chemical Society 73, 3548-3548.

Kendrew, J.C., Bodo, G., Dintzis, H.M., Parrish, R.G., Wyckoff, H., and Phillips, D.C. (1958). A three-dimensional model of the myoglobin molecule obtained by x- ray analysis. Nature 181, 662-666.

Woolley, D.W. (1959). Antimetabolites. They help in discovery of metabolic pathways and in the understanding and treatment of some diseases 129, 615-621.

Butenandt, A. (1960). Zur Geschichte der Sterin- und Vitamin-Forschung. Adolf Windaus zum Gedächtnis. Angewandte Chemie 72, 645-651.

Dalgliesh, C.E., Horning, E.C., Horning, M.G., Knox, K.L., and Yarger, K. (1966). A gas-liquid-chromatographic procedure for separating a wide range of metabolites occuring in urine or tissue extracts. Biochem J 101, 792-810.

Arpino, P., Baldwin, M.A., and McLafferty, F.W. (1974). Liquid chromatography- mass spectrometry. II—continuous monitoring. Biomedical Mass Spectrometry 1, 80-82.

Fenn, J., Mann, M., Meng, C., Wong, S., and Whitehouse, C. (1989). Electrospray ionization for mass spectrometry of large biomolecules. Science (New York, N.Y.) 246, 64-71.

Lerner, R.A., Siuzdak, G., Prospero-Garcia, O., Henriksen, S.J., Boger, D.L., and Cravatt, B.F. (1994). Cerebrodiene: a brain lipid isolated from sleep-deprived cats. Proceedings of the National Academy of Sciences 91, 9505-9508.

Arpino, P., ed. (2006). History of LC-MS development and interfacing. (Elsevier).

1. A History of Mass Spectrometry and Metabolomics

Beckonert, O., Keun, H.C., Ebbels, T.M., Bundy, J., Holmes, E., Lindon, J.C., and Nicholson, J.K. (2007). Metabolic profiling, metabolomic and metabonomic procedures for NMR spectroscopy of urine, plasma, serum and tissue extracts. Nature Protocols 2, 2692-2703.

Bergstrom, S., and Samuelsson, B. (1965). Prostaglandins. Annual Review of Biochemistry 34, 101-108.

Buchner, E. (1897). Alkoholische Gährung ohne Hefezellen. Berichte der deutschen chemischen Gesellschaft 30, 117-124.

Cravatt, B.F., Prospero-Garcia, O., Siuzdak, G., Gilula, N.B., Henriksen, S.J., Boger, D.L., and Lerner, R.A. (1995). Chemical characterization of a family of brain lipids that induce sleep. Science (New York, N.Y.) 268, 1506-1509.

Dagley, S., and Nicholson, D.E. (1970). An Introduction to Metabolic Pathways. (Blackwell Scientific Publications Ltd.).

Ettre, L.S., and Sakodynskii, K.I. (1993). M. S. Tswett and the discovery of chromatography II: Completion of the development of chromatography (1903– 1910). Chromatographia 35, 329-338.

Frainay, C., Schymanski, E.L., Neumann, S., Merlet, B., Salek, R.M., Jourdan, F., and Yanes, O. (2018). Mind the Gap: Mapping Mass Spectral Databases in Genome-Scale Metabolic Networks Reveals Poorly Covered Areas. Metabolites 8, 51.

Freiherr von Liebig, J. (1843). Die Thier-Chemie, oder, Die organische Chemie in ihrer Anwendung auf Physiologie und Pathologie. (Braunschweig: F. Vieweg und Sohn).

Garrod, A. (1902). THE INCIDENCE OF ALKAPTONURIA : A STUDY IN CHEMICAL INDIVIDUALITY. The Lancet 160, 1616-1620.

Gates, S.C., and Sweeley, C.C. (1978). Quantitative metabolic profiling based on gas chromatography. Clinical Chemistry 24, 1663-1673.

Gohlke, R.S. (1959). Time-of-Flight Mass Spectrometry and Gas-Liquid Partition Chromatography. Analytical Chemistry 31, 535-541.

Guijas, C., Montenegro-Burke, J.R., Domingo-Almenara, X., Palermo, A., Warth, B., Hermann, G., Koellensperger, G., Huan, T., Uritboonthai, W., Aisporna, A.E., et al. (2018). METLIN: A Technology Platform for Identifying Knowns and Unknowns. Analytical Chemistry 90, 3156-3164.

Haddon, W.F., and McLafferty, F.W. (1968). Metastable ion characteristics. VII. Collision-induced metastables. Journal of the American Chemical Society 90, 4745-4746.

Hoult, D.I., Busby, S.J.W., Gadian, D.G., Radda, G.K., Richards, R.E., and Seeley, P.J. (1974). Observation of tissue metabolites using 31P nuclear magnetic resonance. Nature 252, 285-287.

King, Z.A., Lu, J., Dräger, A., Miller, P., Federowicz, S., Lerman, J.A., Ebrahim, A., Palsson, B.O., and Lewis, N.E. (2016). BiGG Models: A platform for integrating, standardizing and sharing genome-scale models. Nucleic Acids Res 44, D515- 522.

Knoop, F. (1904). Der Abbau aromatischer fettsäuren im tierkörper. Beitraege zur Chemischen Physiologie und Pathologie 6, 150-162.

Lipmann, F., and Kaplan, N.O. (1946). A COMMON FACTOR IN THE ENZYMATIC ACETYLATION OF SULFANILAMIDE AND OF CHOLINE. Journal of Biological Chemistry 162, 743-744.

Marquis, E. (1896). Über den Verbleib des Morphin im tierischen Organismus. In Pharmazeutische Zentralhalle für Deutschland (Jurjew, Arb.: Der Pharm. Inst. zu Dorpat), p. 117.

Martin, A.J., and Synge, R.L. (1941). A new form of chromatogram employing two liquid phases: A theory of chromatography. 2. Application to the micro- determination of the higher monoamino-acids in proteins. The Biochemical journal 35, 1358-1368.

Purcell, E.M., Torrey, H.C., and Pound, R.V. (1946). Resonance Absorption by Nuclear Magnetic Moments in a Solid. Physical Review 69, 37-38.

Rabi, I.I., Zacharias, J.R., Millman, S., and Kusch, P. (1938). A New Method of Measuring Nuclear Magnetic Moment. Physical Review 53, 318-318.

Reinitzer, F. (1888). Beiträge zur Kenntniss des Cholesterins. Monatshefte für Chemie und verwandte Teile anderer Wissenschaften 9, 421-441.

Reinitzer, F. (1989). Contributions to the knowledge of cholesterol. Liquid Crystals 5, 7-18.

Rinschen, M.M., Ivanisevic, J., Giera, M., and Siuzdak, G. (2019). Identification of bioactive metabolites using activity metabolomics. Nature reviews. Molecular cell biology 20, 353-367.

Röntgen, W.C. (1898). Ueber eine neue Art von Strahlen. Annalen der Physik 300, 12-17.

Röst, H.L., Sachsenberg, T., Aiche, S., Bielow, C., Weisser, H., Aicheler, F., Andreotti, S., Ehrlich, H.C., Gutenbrunner, P., Kenar, E., et al. (2016). OpenMS: a flexible open-source software platform for mass spectrometry data analysis. Nature methods 13, 741-748.

Schoenheimer, R., and Rittenberg, D. (1935). DEUTERIUM AS AN INDICATOR IN THE STUDY OF INTERMEDIARY METABOLISM. I. Journal of Biological Chemistry 111, 163-168.

Smith, C.A., O'Maille, G., Want, E.J., Qin, C., Trauger, S.A., Brandon, T.R., Custodio, D.E., Abagyan, R., and Siuzdak, G. (2005). METLIN: a metabolite mass spectral database. Therapeutic drug monitoring 27, 747-751.

Smith, C.A., Want, E.J., O'Maille, G., Abagyan, R., and Siuzdak, G. (2006). XCMS: processing mass spectrometry data for metabolite profiling using nonlinear peak alignment, matching, and identification. Analytical Chemistry 78, 779-787.

Teranishi, R., Mon, T.R., Robinson, A.B., Cary, P., and Pauling, L. (1972). Gas chromatography of volatiles from breath and urine. Analytical Chemistry 44, 18-20.

Thaulow, M.C.J. (1838). Ueber die Zuckersäure. Annalen der Pharmacie 27, 113-130.

Tollens, B. (1882). Ueber ammon-alkalische Silberlösung als Reagens auf Aldehyd. Berichte der deutschen chemischen Gesellschaft 15, 1635-1639.

Williams, R.J. (1956). Biochemical Individuality. The Basis for the Genetotrophic Concept. (New York: John Willey & Sons).

Chapter 2
Ionization

Mass Spectrometry

Mass spectrometry has been described as the smallest scale in the world, not because of the mass spectrometer's size but because of the size of what it weighs -- molecules. Over the past three decades mass spectrometry has undergone tremendous technological improvements allowing for its application to proteins, peptides, carbohydrates, DNA, drugs, and many other biologically relevant molecules. Due to ionization sources such as electrospray ionization (ESI) and matrix-assisted laser desorption/ionization (MALDI), mass spectrometry has become an irreplaceable tool in the biological sciences. This chapter provides an overview of mass spectrometry ionization sources and their significance in the development of mass spectrometry in biomolecular analysis.

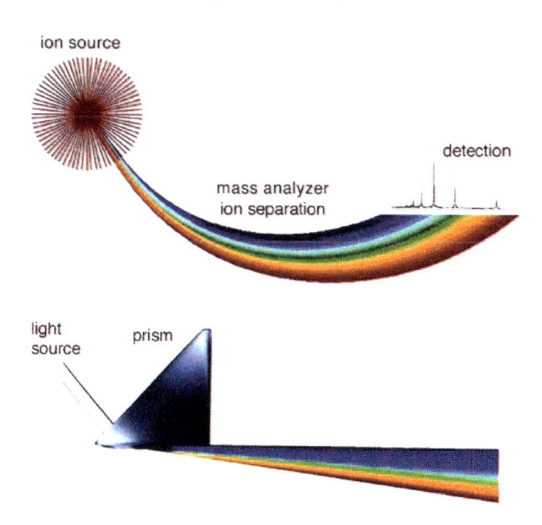

Figure 2.1. Mass analysis process as compared to the dispersion of light.

What is Mass Spectrometry?

A mass spectrometer determines the mass of a molecule by measuring the mass-to-charge ratio (m/z) of its ion. Ions are generated by inducing either the loss or gain of a charge from a neutral species. Once formed, ions are electrostatically directed into a mass analyzer where they are separated according to *m/z* and finally detected. The result of molecular ionization, ion separation, and ion detection is a spectrum that can provide molecular mass and even structural information. An analogy can be drawn between a mass spectrometer and a prism, as shown in **Figure 2.1**. In the prism, light is separated into its component wavelengths which are then detected with an optical receptor, such as visualization. Similarly, a mass spectrometer's ions are separated in the mass analyzer, digitized, and detected by an ion detector (such as an electron multiplier, **Chapter 3**).

John B. Fenn, the originator of electrospray ionization for biomolecules and the 2002 Nobel Laureate in Chemistry, probably gave the best answers to the question "what is mass spectrometry?":

Mass spectrometry is the art of measuring atoms and molecules to determine their molecular weight. Such mass or weight information is sometimes sufficient, frequently necessary, and always useful in determining the identity of a species. To practice this art one puts charge on the molecules of interest, i.e., the analyte, then measures how the trajectories of the resulting ions respond in vacuum to various combinations of electric and magnetic fields. Clearly, the sine qua non of such a method is the conversion of neutral analyte molecules into ions. For small and simple species the ionization is readily carried by gas-phase encounters between the neutral molecules and electrons, photons, or other ions. In recent years, the efforts of many investigators have led to new techniques for producing ions of species too large and complex to be vaporized without substantial, even catastrophic, decomposition.

2. Ionization

Some Basics

Four basic components are, for the most part, standard in all mass spectrometers (**Figure 2.2**): a sample inlet, an ionization source, a mass analyzer, and an ion detector. Some instruments combine the sample inlet and the ionization source, while others combine the mass analyzer and the detector. However, all sample molecules undergo the same processes regardless of instrument configuration. Sample molecules are introduced into the instrument through a sample inlet. Once inside the instrument, the sample molecules are converted to ions in the ionization source, before being electrostatically propelled into the mass analyzer. Ions are then separated according to their m/z within the mass analyzer. The detector converts the ion energy into electrical signals, which are then transmitted to a computer.

Sample Introduction Techniques

Sample introduction was an early challenge in mass spectrometry. To perform mass analysis on a sample, which is initially at atmospheric pressure (760 torr), it must be introduced into the instrument in such a way that the vacuum inside the instrument remains relatively unchanged (~10^{-6} torr). The most common methods of sample introduction are direct insertion with a probe or plate commonly used with MALDI-MS, direct infusion, or injection into the ionization source such as ESI-MS.

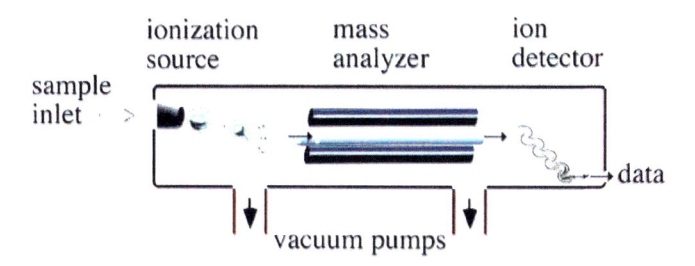

Figure 2.2. Components of a mass spectrometer. Note that the ion source does not have to be within the vacuum, for example, ESI and

APCI are at atmospheric pressure and are known as atmospheric pressure ionization (API) sources.

Direct Insertion: Using an insertion probe/plate (**Figure 2.3**) is a very simple way to introduce a sample into an instrument. The sample is first placed onto a probe and then inserted into the ionization region of the mass spectrometer, typically through a vacuum interlock. The sample is then subjected to any number of desorption processes, such as laser desorption or direct heating, to facilitate vaporization and ionization.

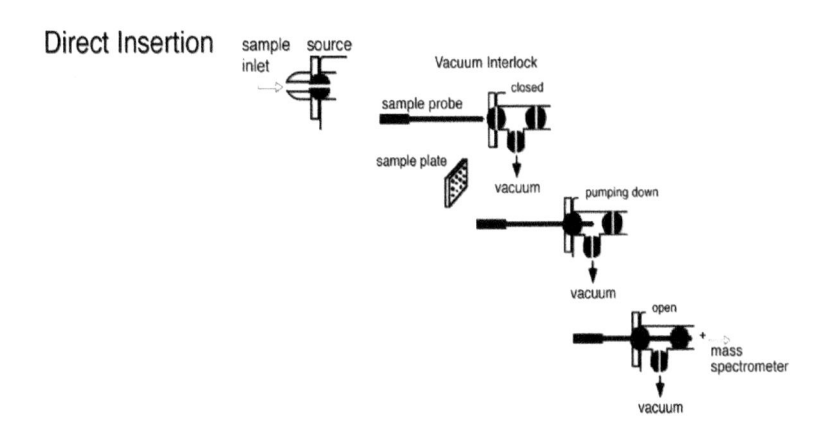

Figure 2.3. Samples can be introduced into the mass spectrometer using a direct insertion probe, a capillary column (EI with GC/MS or ESI) or a sample plate (MALDI). The vacuum interlock allows for the vacuum of the mass spectrometer to be maintained while the instrument is not in use.

Direct Infusion: A simple capillary or a capillary column is typically used to introduce a sample as a gas or in solution. Direct infusion is also useful because it can efficiently introduce small quantities of sample into a mass spectrometer without compromising the vacuum. Capillary columns are routinely used to interface separation techniques with the ionization source of a mass spectrometer. These techniques, including gas chromatography (GC) and liquid chromatography (LC), also serve to separate a solution's different components prior to mass analysis. In gas

2. Ionization

chromatography, separation of different components occurs within a glass capillary column. As the vaporized sample exits the gas chromatograph, it is directly introduced into the mass spectrometer.

In the 1980s the incapability of liquid chromatography (LC) with mass spectrometry was due largely to the ionization techniques being unable to handle the continuous flow of LC. However, electrospray ionization (ESI), atmospheric pressure chemical ionization (APCI) and atmospheric pressure photoionization (APPI) now allows LC/MS to be performed routinely (**Figure 2.4**).

Liquid Chromatography Mass Spectrometry

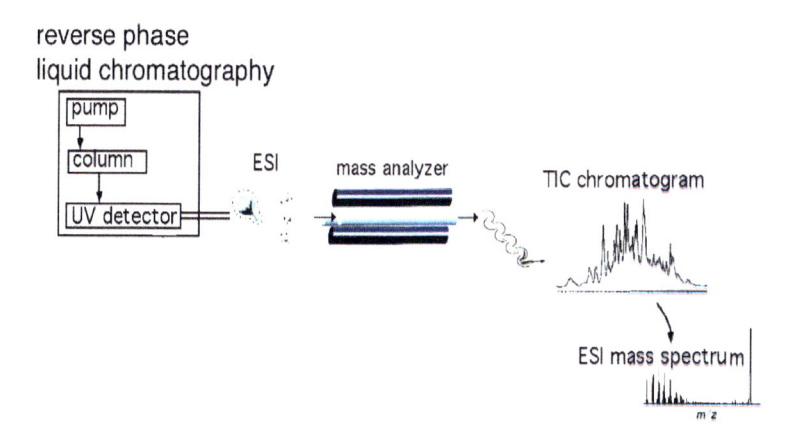

Figure 2.4. Interfacing liquid chromatography with electrospray ionization mass spectrometry. Liquid chromatography/mass spectrometry (LC/MS) ion chromatogram and the corresponding electrospray mass spectrum are shown. Gas chromatography mass spectrometry (GC/MS) produces results in much the same way as LC/MS, however, GC/MS uses an electron ionization source, which is limited by thermal vaporization (UV refers to ultraviolet and TIC is the total ion current).

Ionization

Ionization method refers to the mechanism of ionization while the ionization source is the mechanical device that allows ionization to occur. The different ionization methods, summarized here, work by either ionizing a neutral molecule through electron ejection, electron capture, protonation, cationization, or deprotonation, or by transferring a charged molecule from a condensed phase to the gas phase.

Protonation

Scheme 2.1. An example of a mass spectrum obtained via protonation.

$$M + H^+ \rightarrow MH^+$$

$$H_2N\text{-RGASRR-OH} + H^+$$

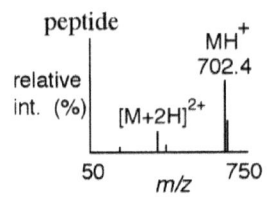

peptide spectrum: relative int. (%) vs m/z; peaks at [M+2H]$^{2+}$ and MH$^+$ 702.4; axis 50 to 750

Protonation is a method of ionization by which a proton is added to a molecule, producing a net positive charge of 1+ for every proton added. Positive charges tend to reside on the more basic residues of the molecule, such as amines, to form stable cations. Peptides are often ionized via protonation. Protonation can be achieved via matrix-assisted laser desorption/-ionization (MALDI), electrospray ionization (ESI) and atmospheric pressure chemical ionization (APCI).

Deprotonation

Scheme 2.2. An example of a mass spectrum of sialic acid obtained via deprotonation.

$$M - H^+ \rightarrow [\ -H]^-$$

sialic acid structure; spectrum: relative int. (%) vs m/z; [M−H]$^-$ 308.1; axis 10 to 330

Deprotonation is an ionization method by which the net negative charge of 1- is achieved through the removal of a proton from a molecule. This mechanism of ionization, commonly achieved via MALDI, ESI, and APCI is very useful for acidic species including phenols, carboxylic acids, and sulfonic acids. The negative ion mass spectrum of sialic acid is shown in **Scheme 2.2**.

2. Ionization

Cationization

Scheme 2.3. An example of a mass spectrum obtained via cationization.

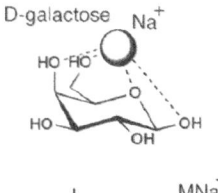

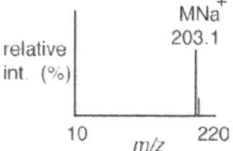

M + Cation⁺ →
Cation⁺

D-galactose Na⁺

MNa⁺
203.1

relative int. (%)

10   220
m/z

Cationization is a method of ionization that produces a charged complex by non-covalently adding a positively charged ion to a neutral molecule. While protonation could fall under this same definition, cationization is distinct for its addition of a cation adduct other than a proton (e.g. alkali, ammonium). Moreover, it is known to be useful with molecules unstable to protonation. The binding of cations other than protons to a molecule is naturally less covalent, therefore, the charge remains localized on the cation. This minimizes delocalization of the charge and fragmentation of the molecule. Cationization is commonly achieved via MALDI, ESI, and APCI. Carbohydrates are excellent candidates for this ionization mechanism, with Na⁺ a common cation adduct.

Transfer of a charged molecule to the gas phase

Scheme 2.4. An example of a mass spectrum of tetraphenylphosphine obtained via transfer of a charged species from solution into the gas phase.

M⁺ solution → M⁺ gas

tetraphenylphosphine

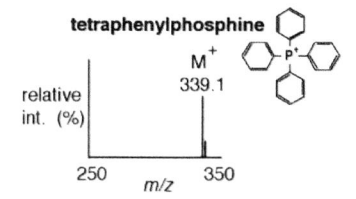

M⁺
339.1

relative int. (%)

250   350
m/z

The transfer of compounds already charged in solution is normally achieved through the desorption or ejection of the charged species from the condensed phase into the gas phase. This transfer is commonly achieved via MALDI or ESI. The positive ion mass spectrum of tetraphenylphosphine is shown in **Scheme 2.4.**

Ionization

Electron ejection

Scheme 2.5. An example of a mass spectrum obtained via electron ejection.

$M \rightarrow M^{+\cdot}$

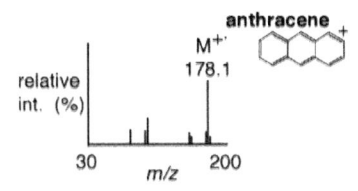

As its name implies, electron ejection achieves ionization through the ejection of an electron to produce a 1+ net positive charge, often forming radical cations. Observed most commonly with electron ionization (EI) sources, electron ejection is usually performed on relatively nonpolar compounds with low molecular weights, and it is also known to generate significant fragment ions. The mass spectrum resulting from electron ejection of anthracene is shown in **Scheme 2.5**.

Electron capture

Scheme 2.6. An example of a mass spectrum obtained via electron capture. Electron capture is commonly achieved via electron ionization (EI).

$M \rightarrow M^{-\cdot}$

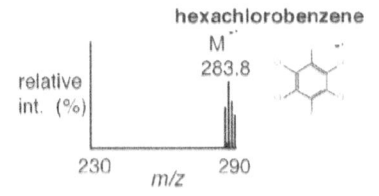

With the electron capture ionization method, a net negative charge of 1- is achieved with the absorption or capture of an electron. It is a mechanism of ion-ization primarily observed for molecules with a high electron affinity, such as halogenated compounds. The electron capture mass spectrum of hexachloro-benzene is shown in **Scheme 2.6**.

Table 2.1. Ionization methods, advantages & disadvantages.

Ionization Method	Advantages	Disadvantages
Protonation (Positive ions)	— many compounds will accept a proton to become ionized — many ionization sources (ESI, APCI & MALDI) facilitate protonation	— some compounds are not stable to protonation (e.g., carbohydrates) or cannot accept a proton (e.g., hydrocarbons)
Cationization (Positive ions)	— many compounds will accept a cation, such as Na^+ or K^+ — many ionization sources (ESI, APCI & MALDI) facilitate cationization	— tandem mass spectrometry experiments on cationized molecules will often generate limited or no fragmentation information
Deprotonation (Negative ions)	— useful for compounds that are somewhat acidic — many ionization sources (ESI, APCI & MALDI) facilitate deprotonation	— compound specific
Transfer of charged molecule to gas phase (Positive or negative ions)	— useful when compound is already charged — many ionization sources (ESI, APCI & MALDI) facilitate this	— only useful for pre-charged ions
Electronejection (Positive ions)	— observed with electron ionization (EI) and can provide molecular mass as well as fragmentation information	— often generates too much fragmentation — it can be unclear whether the highest mass ion is the molecular ion or a fragment — molecules subjected to thermal degradation
Electroncapture (Negative ions)	— observed with electron ionization (EI) and can provide molecular mass as well as fragmentation information	— often generates too much fragmentation — unclear whether the highest mass ion is the molecular ion or fragment — molecules subjected to thermal degradation

Ionization Sources

Prior to the 1980s, electron ionization (EI) was the primary ionization source for mass analysis. However, EI limited chemists and biochemists to small molecules well below the mass range of common bio-organic compounds. This limitation motivated scientists such as John B. Fenn, Koichi Tanaka, Franz Hillenkamp, Michael Karas, Graham Cooks, and Michael Barber to develop the new generation of ionization techniques, including fast atom/ion bombardment (FAB), matrix-assisted laser desorption/ionization (MALDI), and electrospray ionization (ESI) (Table 2.2). These techniques have revolutionized biomolecular analyses, especially for large molecules. Among them, ESI and MALDI have clearly evolved to be the methods of choice when it comes to biomolecular analysis.

Table 2.2. Ionization Sources

Ionization Source	Acronym	Event
Electrospray ionization	ESI	evaporation of charged droplets
Nanoelectrospray ionization	nanoESI	evaporation of charged droplets
Atmospheric pressure chemical ionization	APCI	corona discharge and proton transfer
Matrix-assisted laser desorption/ionization	MALDI	photonabsorption/proton transfer
Desorption/ionization on silicon	DIOS	photonabsorption/proton transfer
Fast atom/ion bombardment	FAB	ion desorption/proton transfer
Electron ionization	EI	electron beam/electron transfer
Chemical ionization	CI	proton transfer

2. Ionization

MALDI and ESI are now the most common ionization sources for biomolecular mass spectrometry, offering excellent mass range and sensitivity (**Figure 2.5**). The following section will focus on the principles of ionization sources, providing some details on the practical aspects of their use as well as ionization mechanisms.

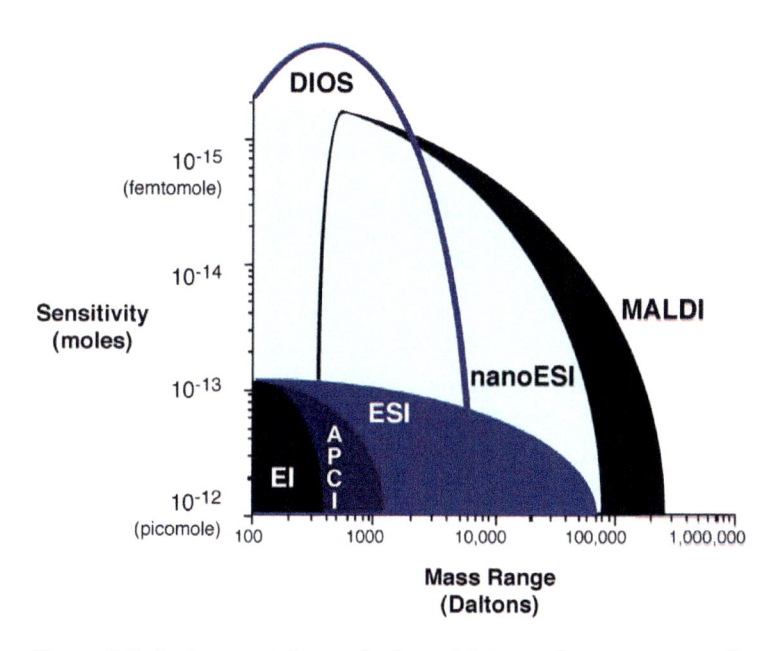

Figure 2.5. A glance at the typical sensitivity and mass ranges allowed by different ionization techniques provides a clear answer to the question of which are most useful; electrospray ionization (ESI), nanoelectrospray ionization (nanoESI), and matrix-assisted laser desorption ionization (MALDI) have a high sensitivity and mass range.

Electrospray Ionization

The original idea of electrospray developed in the 1930s and 1960s, was rejuvenated with its application to biomolecules in the 1980s resulting in the 2002 Nobel Prize for John Fenn. The first electrospray experiments were carried out by Chapman in the late 1930s and the practical development of electrospray ionization for mass spectrometry was accomplished by Dole in the late 1960s. Dole also discovered the important phenomenon of multiple charging of molecules. It was Fenn's work that ultimately led to the modern-day technique of electrospray ionization mass spectrometry and its application to biological macromolecules.

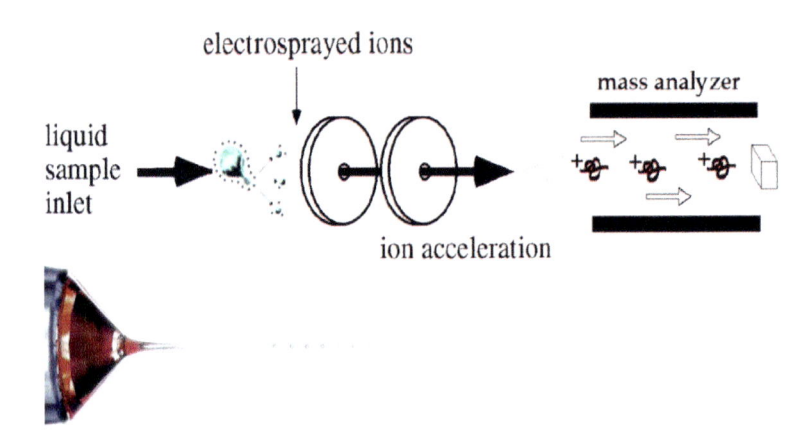

Figure 2.6. Electrospray ionization (ESI). Credit lower image to Fossilionsource.

Electrospray ionization (ESI) is a method routinely used with endogenous metabolites, peptides, proteins, carbohydrates, small oligonucleotides, synthetic polymers, and lipids. ESI produces gaseous ionized molecules directly from a liquid solution. It operates by creating a fine spray of highly charged droplets in the presence of an electric field. (An illustration of the electrospray ionization process is shown in Figures 2.6 and 2.7). The sample solution is sprayed from a region of the strong electric field at the tip of a metal nozzle maintained at a potential of anywhere from 700 V to 5000 V. The nozzle (or needle) to which the potential is applied serves to disperse the solution into a fine spray of charged droplets. Either dry gas, heat, or both are applied to the droplets at atmospheric pressure thus causing the solvent to evaporate from each droplet. As the size of the charged droplet decreases, the charge density on its surface increases. The mutual Coulombic repulsion between like charges on this surface becomes so great that it exceeds the forces of surface tension, and ions are ejected from the droplet through a "Taylor

2. Ionization

cone" Figure 2.7. Another possibility is that the droplet explodes releasing the ions. In either case, the emerging ions are directed into an orifice through electrostatic lenses leading to the vacuum of the mass analyzer. Because ESI involves the continuous introduction of solution, it is particularly suitable to interface with HPLC or capillary electrophoresis.

Charged Electrospray Droplet

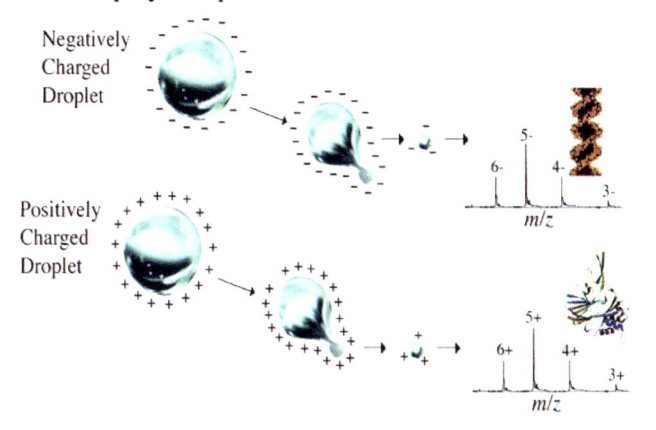

Figure 2.7. Ion formation from electrospray ionization source. The electrospray ionization source uses a stream of air or nitrogen, heat, a vacuum, or a solvent sheath (e.g., methanol) to facilitate droplet desolvation. Ion ejection occurs through a "Taylor cone" (central droplet) where ions are then electrostatically directed into the analyzer.

Electrospray ionization is conducive to the formation of singly charged small molecules but is also well-known for producing multiply charged species of larger molecules. This is an important phenomenon because the mass spectrometer measures the mass-to-charge ratio (m/z) and therefore multiple charging makes it possible to observe very large molecules with an instrument having a relatively small mass range. Fortunately, the software available with all electrospray mass spectrometers facilitates the molecular weight calculations necessary to determine the actual mass of the multiply charged species.

Ionization Sources

Figures 2.7-2,9 illustrate the different charge states on two different proteins, where each of the peaks in the mass spectra can be associated with different charge states of the molecular ion. Multiple charging has other important advantages in tandem mass spectrometry. One advantage is that upon fragmentation you observe more fragment ions with multiply charged precursor ions than with singly charged precursor ions.

Multiple charging: A 10,000 Da protein and its theoretical mass spectrum with up to five charges are shown in **Figure 2.8**. The mass of the protein remains the same, yet the m/z ratio varies depending upon the number of charges on the protein. Protein ionization is usually the result of protonation, which not only adds charge but also increases the mass of the protein by the number of protons added. This effect on the m/z applies equally for any mechanism of molecular ionization resulting in a positively or negatively charged molecular ion, including the addition or ejection of charge-carrying species other than protons (e.g. Na+ and Cs+). Multiple positive charges are observed for proteins, while for oligonucleotides negative charging (with ESI) is typical.

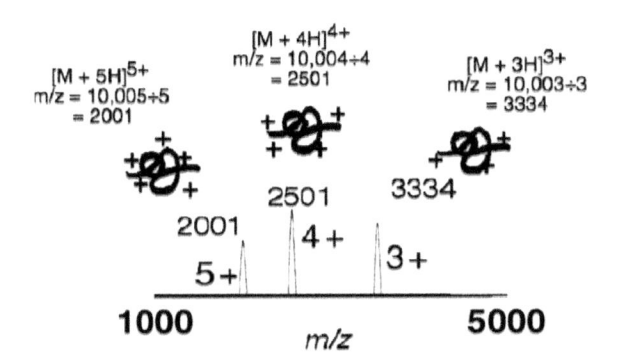

Figure 2.8. A theoretical protein with a molecular weight of 10,000 generates three different peaks with the ions containing 5, 4, and 3 charges, respectively. The mass spectrometer detects each of the protein ions at 2001, 2501, and 3334, respectively.

2. Ionization

Although electrospray mass spectrometers are equipped with software that will calculate molecular weight, an understanding of how the computer makes such calculations from multiply-charged ions is beneficial. Equations 1.1 - 1.5 and Figure 1.9 offer a simple explanation, where we assume p1 and p2 are adjacent peaks and differ by a single charge, which is equivalent to the addition of a single proton.

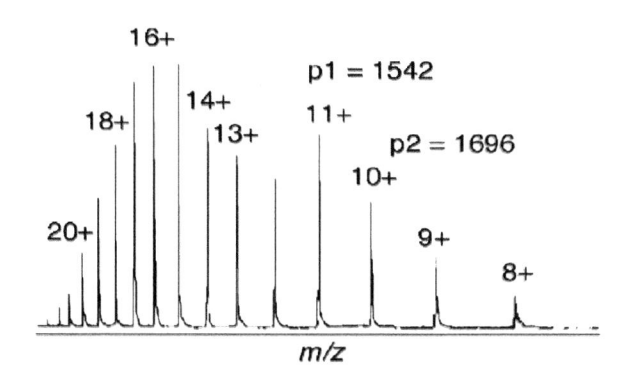

Figure 2.9. The multiply charged ions of myoglobin generated from ESI. The different peaks represent different charge states of myoglobin. The molecular weight can be determined using Equations 1.1 - 1.3.

$$p = m/z \qquad (2.1)$$

$$p_1 = (M_r + z_1)/z_1 \qquad (2.2)$$

$$p_2 = \{M_r + (z_1 - 1)\}/(z_1 - 1) \qquad (2.3)$$

p = a peak in the mass spectrum
m = total mass of an ion
z = total charge
M_r = average mass of protein

p_1 = m/z value for p1
p_2 = m/z value for p2
z_1 = charge on peak p1

Equations 2.2 and 2.3 can be solved for the two unknowns, M_r and z_1.

For the peaks in the mass spectrum of myoglobin shown in Figure 1.9, p1=1542, and p2=1696.

$$1542 z_1 = M_r + z_1 \tag{2.4}$$

$$1696 (z_1 - 1) = M_r + (z_1 - 1) \tag{2.5}$$

Solving the two equations: $M_r = 16,951$ Da for $z_1 = 11$

Table 2.3. Advantages and disadvantages of ESI.

Advantages	Disadvantages
— practical mass range of up to 70,000 Da	— the presence of salts and ion-pairing agents like TFA can reduce sensitivity.
— good sensitivity with femtomole to low picomole sensitivity typical	— complex mixtures can reduce sensitivity.
— soft ionization method, capable of producing noncovalent complexes	— simultaneous mixture analysis can be poor.
— easily adaptable to liquid chromatography	— multiple charging can be confusing especially in mixture analysis.
— easily adaptable to tandem MS	— sample purity is important.
— multiple charging allows for analysis of high mass ions	— carryover from sample to sample.
— no matrix interference	

Electrospray Solvents

Many solvents can be used in ESI and are chosen based on the solubility of the compound of interest, the volatility of the solvent and the solvent's ability to donate a proton. Typically, protic primary solvents such as methanol, 50/50 methanol/water, or 50/50 acetonitrile/H2O are used, while aprotic cosolvents, such as 10% DMSO in water, as well as isopropyl alcohol are used to improve solubility for some compounds. Although 100% water is used in ESI, water's relatively low vapor pressure has a detrimental effect on sensitivity; better sensitivity is obtained when a volatile organic solvent is added. Some compounds require the use of straight chloroform with 0.1% formic acid added to facilitate ionization. This approach, while less sensitive, can be effective for otherwise insoluble compounds. Buffers such as Na+, K+ phosphate, and salts present a problem for ESI by lowering the

2. Ionization

vapor pressure of the droplets resulting in reduced signal through an increase in droplet surface tension resulting in a reduction of volatility (see **Chapter 3** for quantitative information on salt effects). Consequently, volatile buffers such as ammonium acetate are more effective.

Configuration of the Electrospray Ion Source

The off-axis ESI configuration now used in many instruments to introduce the ions into the analyzers (as shown in **Figure 2.10**) has turned out to be very valuable for high flow rate applications. The primary advantage of this configuration is that the flow rates can be increased without contaminating or clogging the inlet. Off-axis spraying is important because the entrance to the analyzer is no longer saturated by solvent, thus keeping droplets from entering and contaminating the inlet. Instead, only ions are directed toward the inlet. This makes ESI even more compatible with LC/MS at the milliliter per minute flow rates.

Figure 2.10. An example of off-axis ESI.

Nanoelectrospray Ionization (NanoESI)

Low flow electrospray, originally described by Wilm and Mann, has been called nanoelectrospray, nanospray, and micro-electrospray. This ionization source is a variation on ESI, where the spray needle has been made very small and is positioned close to the entrance to the mass analyzer (**Figure 2.11**). The end result of this rather simple adjustment is increased efficiency, which includes a reduction in the amount of sample needed.

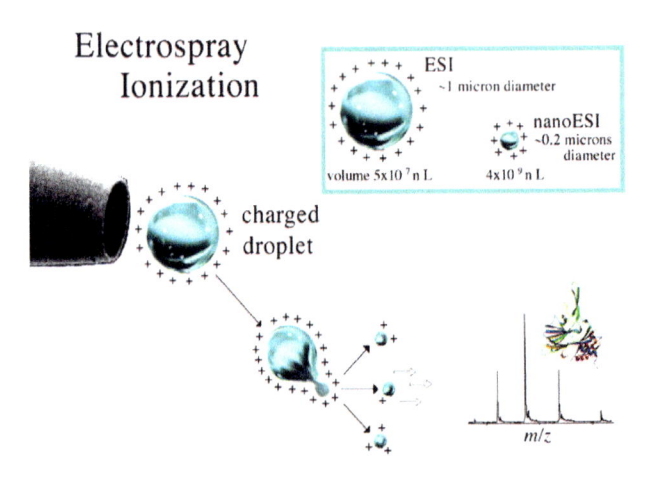

Figure 2.11. Nanoelectrospray ionization (NanoESI).

The flow rates for nanoESI sources are on the order of tens to hundreds of nanoliters per minute. In order to obtain these low flow rates, nanoESI uses emitters of pulled and in some cases metallized glass or fused silica that have a small orifice (~5µ). The dissolved sample is added to the emitter and a pressure of ~30 PSI is applied to the back of the emitter. Effusing the sample at very low flow rates allows for high sensitivity. Also, the emitters are positioned very close to the entrance of the mass analyzer, therefore ion transmission to the mass analyzer is much more efficient. For instance, the analysis of a 5 mM solution of a peptide by nanoESI would be performed in 1 minute, consuming ~50 femtomoles of sample. The same experiment performed with normal ESI in the same time period would require 5 picomoles, or 100 times more sample than for nanoESI. In addition, since the droplets are typically smaller with nanoESI than normal ESI (**Figure 2.11**), the amount of evaporation necessary to obtain ion

2. Ionization

Atmospheric Pressure Chemical Ionization

APCI has also become an important ionization source because it generates ions directly from solution and it can analyze relatively nonpolar compounds. Like electrospray, the liquid effluent of APCI (**Figure 2.12**) is introduced directly into the ionization source. However, the similarity stops there. The droplets are not charged and the APCI source contains a heated vaporizer, which facilitates rapid desolvation/vaporization of the droplets. Vaporized sample molecules are carried through an ion- molecule reaction region at atmospheric pressure. The ionization originates from the solvent being excited/ionized from the corona discharge. Because the solvent ions are present at atmospheric pressure conditions, chemical ionization of analyte molecules is very efficient; at atmospheric pressure analyte molecules collide with the reagent ions frequently. Proton transfer (for protonation MH+ reactions) occurs in the positive mode, and either electron transfer or proton loss, ([M-H]-) in the negative mode. The moderating influence of the solvent clusters on the reagent ions, and of the high gas pressure, reduces fragmentation during ionization and results in primarily intact molecular ions. Multiple charging is typically not observed presumably because the ionization process is more energetic than ESI.

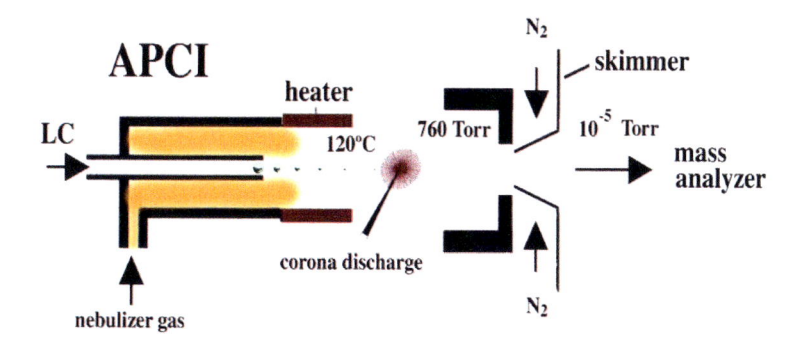

Figure 2.12. Atmospheric pressure chemical ionization (APCI) mass spectrometry.

Atmospheric Pressure Photoionization

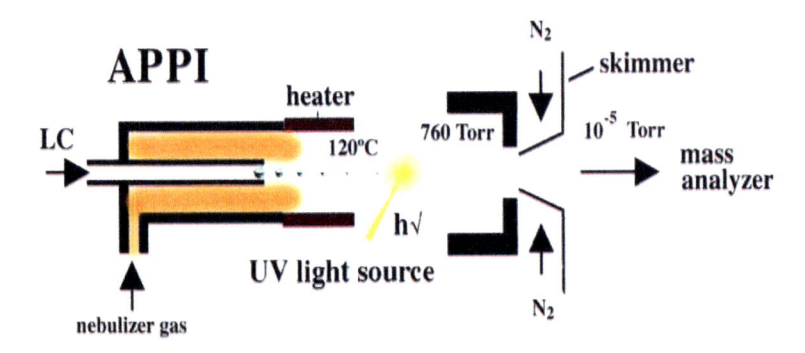

Figure 2.13. Atmospheric pressure photoionization (APPI) mass spectrometry.

Atmospheric pressure photoionization (APPI) has recently become an important ionization source because it generates ions directly from solution with relatively low background and is capable of analyzing relatively nonpolar compounds. Like APCI, the liquid effluent of APPI (**Figure 2.13**) is introduced directly into the ionization source. The primary difference between APCI and APPI is that the APPI vaporized sample passes through ultra-violet light (a typical krypton light source emits at 10.0 eV and 10.6 eV). Often, APPI is much more sensitive than ESI or APCI and has been shown to have higher signal-to-noise ratios because of lower background ionization. Lower background signal is largely due to high ionization potential of standard solvents such as methanol and water (IP 10.85 and 12.62 eV, respectively) which are not ionized by the krypton lamp.

A disadvantage of both ESI and APCI is that they can generate background ions from solvents. Additionally, ESI is especially susceptible to ion suppression effects, and APCI requires vaporization temperatures ranging from 350-500° C, which can cause thermal degradation.

APPI induces ionization via two different mechanisms. The first is direct photoexcitation, allowing for electron ejection and the generation of the positive ion radical cation (M+.). The

2. Ionization

APPI source imparts light energy that is higher than the ionization potentials (IPs) of most target molecules, but lower than most of the IPs of air and solvent molecules, thus removing them as interferants. In addition, because little excess energy is deposited in the molecules, there is minimal fragmentation.

The second mechanism is atmospheric pressure photo-induced chemical ionization which is like APCI in that it involves charge transfer to produce protonation (MH+) or proton loss ([M- H]-) to generate negative ions.

To initiate chemical ionization, a photo ionizable reagent, also called a dopant, is added to the eluant. Upon photoionization of the dopant, charge transfer occurs to the analyte. Typical dopants in positive mode include acetone and toluene. Acetone also serves as a dopant in negative mode.

The ionization mechanism (M+. versus [M+H]+) that a molecule undergoes depends on the proton affinity of the analyte, the solvent, and the type of dopant used.

Matrix-Assisted Laser Desorption/Ionization (MALDI)

Matrix-assisted laser desorption/ionization mass spectrometry (MALDI-MS) was first introduced in 1988 by Tanaka, Karas, and Hillenkamp. It has since become a widespread analytical tool for peptides, proteins, and most other biomolecules (oligonucleotides, carbohydrates, natural products, and lipids). The efficient and directed energy transfer during a matrix-assisted laser-induced desorption event provides high ion yields of the intact analyte and allows for the measurement of compounds with sub-picomole sensitivity. In addition, the utility of MALDI for the analysis of heterogeneous samples makes it very attractive for the mass analysis of complex biological samples such as proteolytic digests.

Ionization Sources

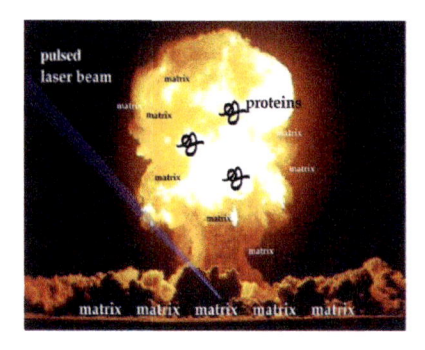

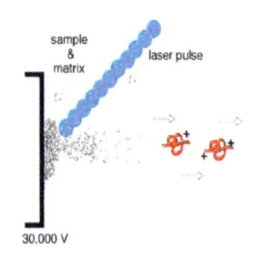

Figure 2.14. The directed energy transfer of the UV laser pulse during a MALDI event allows for relatively small quantities of sample (femtomole to picomole) to be analyzed. In addition, the utility of MALDI mass spectrometry for the analysis of heterogeneous samples makes it very attractive for the mass analysis of biological samples.

While the exact desorption/ionization mechanism for MALDI is not known, it is generally believed that MALDI causes the ionization and transfer of a sample from the condensed phase to the gas phase via laser excitation and vaporization of the sample matrix (**Figure 2.14**). In MALDI analysis, the analyte is first co- crystallized with a large molar excess of a matrix compound, usually a UV-absorbing weak organic acid. Irradiation of this analyte-matrix mixture by a laser result in the vaporization of the matrix, which carries the analyte with it. The matrix plays a key role in this technique. The co-crystallized sample molecules also vaporize, but without having to directly absorb energy from the laser. Molecules sensitive to the laser light are therefore protected from direct UV laser excitation.

MALDI matrix -- A nonvolatile solid material facilitates the desorption and ionization process by absorbing the laser radiation. As a result, both the matrix and any sample embedded in the matrix are vaporized. The matrix also serves to minimize sample damage from laser radiation by absorbing most of the incident energy.

2. Ionization

Once in the gas phase, the desorbed charged molecules are then directed electrostatically from the MALDI ionization source into the mass analyzer. Time-of-flight (TOF) mass analyzers are often used to separate the ions according to their mass-to-charge ratio (m/z). The pulsed nature of MALDI is directly applicable to TOF analyzers since the ion's initial time-of-flight can be started with each pulse of the laser and completed when the ion reaches the detector.

Several theories have been developed to explain desorption by MALDI. The thermal-spike model proposes that the ejection of intact molecules is attributed to poor vibrational coupling between the matrix and analyte, which minimizes vibrational energy transfer from the matrix to the vibrational modes of the analyte molecule, thereby minimizing fragmentation. The pressure pulse theory proposes that a pressure gradient from the matrix is created normal to the surface and desorption of large molecules is enhanced by momentum transfer from collisions with these fast-moving matrix molecules. It is generally thought that ionization occurs through proton transfer or cationization during the desorption process.

The utility of MALDI for biomolecule analyses lies in its ability to provide molecular weight information on intact molecules. The ability to generate accurate information can be extremely useful for protein identification and characterization. For example, a protein can often be unambiguously identified by the accurate mass analysis of its constituent peptides (produced by either chemical or enzymatic treatment of the sample).

Sample-matrix preparation procedures greatly influence the quality of MALDI mass spectra of peptides/proteins (**Figure 2.15**). Among the variety of reported preparation methods, the dried-droplet method is the most frequently used. In this case, a saturated matrix solution is mixed with the analyte solution, giving a matrix-to-sample ratio of about 5000:1. An aliquot (0.5-2.0 µL) of this mixture is then applied to the sample target where it is allowed to dry. Below is an example of how the dried-droplet method is performed:

- Pipet 0.5 µL of sample to the sample plate.
- Pipet 0.5 µL of matrix to the sample plate.
- Mix the sample and matrix by drawing the combined droplet in and out of the pipette.
- Allow to air dry.

  o For peptides, glycoprotreins, small proteins and most compounds: A saturated solution of α- cyano-4-hydroxycinnamic acid in 50:50 ACN:H₂O with 0.1% TFA.

  o For peptides and proteins and other large molecules: a saturated solution of sinapinic acid in 50:50 ACN:H₂O with 0.1% TFA.

  o For peptides, small proteins and small compounds: a saturated solution of 2,5- dihydroxy benzoic acid (DHB) in 50:50 ACN:H₂O.

α-cyano-4-hydroxycinnamic acid (α-cyano or HCCA)	3,5-dimethoxy-4-hydroxycinnamic acid (sinapinic acid)	2,5-dihyroxy benzoic acid (DHB)
peptides and glycopeptides	peptides and proteins	peptides and small proteins

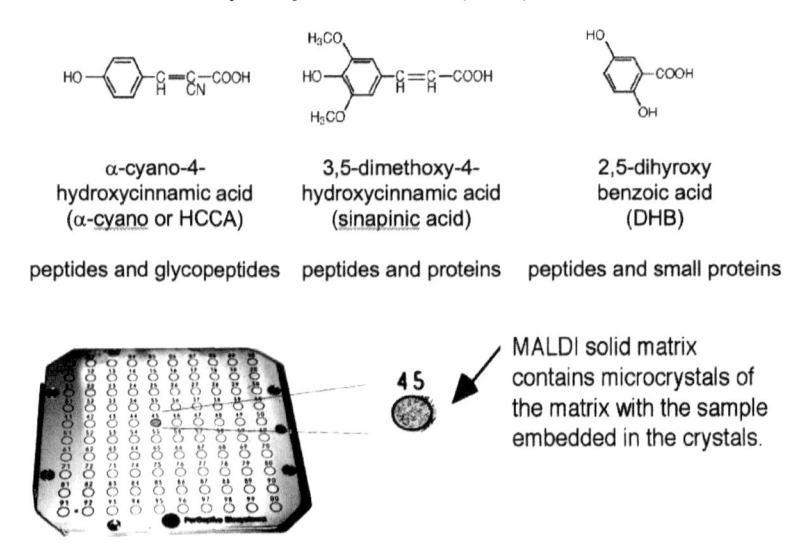

MALDI solid matrix contains microcrystals of the matrix with the sample embedded in the crystals.

Figure 2.15. Commonly used MALDI matrices and a MALDI plate showing the matrix deposition. An advantage of MALDI is that multiple samples can be prepared at the same time, as seen with this multisample plate.

Table 2.4. Advantages and disadvantages of Matrix-Assisted Laser Desorption/Ionization (MALDI).

Advantages	Disadvantages
practical mass range of up to 300,000 Da. Species of much greater mass have been observed using a high current detector.	matrix background, which can be a problem for compounds below a mass of 700 Da. This background interference is highly dependent on the matrix material.
	possibility of photo-degradation by laser desorption/ionization.
typical sensitivity on the order of low femtomole to low picomole. Attomole sensitivity is possible.	acidic matrix used in MALDI may cause degradation on some compounds.
soft ionization with little to no fragmentation observed.	

tolerance of salts in millimolar concentrations.

suitable for the analysis of complex mixtures.

Alternatively, samples can be prepared in a stepwise manner. In the thin layer method, a homogeneous matrix "film" is formed on the target first, and the sample is then applied and absorbed by the matrix. This method yields good sensitivity, resolving power, and mass accuracy. Similarly, in the thick-layer method, nitrocellulose (NC) is used as the matrix additive; once a uniform NC-matrix layer is obtained on the target, the sample is applied. This preparation method suppresses alkali adduct formation and significantly increases the detection sensitivity, especially for peptides and proteins extracted from gels. The sandwich method is another variant in this category. A thin layer of matrix crystals is prepared as in the thin-layer method, followed by the subsequent addition of droplets of (a) aqueous 0.1% TFA, (b) ample and (c) matrix.

Desorption/Ionization on Silicon (DIOS)

DIOS is a matrix-free method that uses pulsed laser desorption/ionization on silicon (**Figure 2.16**). Structured silicon surfaces such as porous silicon or silicon nanowires are UV-absorbing semiconductors with a large surface area (hundreds of m2/cm3). For its application to laser desorption/ionization mass spectrometry, the structure of structured silicon provides a scaffold for retaining solvent and analyte molecules, and the UV absorptivity affords a mechanism for the transfer of the laser energy to the analyte. This fortuitous combination of characteristics allows DIOS to be useful for a large variety of biomolecules including peptides, carbohydrates, and small organic compounds of various types. Unlike other direct, matrix- free desorption techniques, DIOS enables desorption/ionization with little or no analyte degradation.

DIOS has a great deal in common with MALDI. Instrumentation and acquisition using DIOS-MS requires only minor adjustments to the MALDI setup; the chips are simply affixed to a machined MALDI plate and inserted into the spectrometer. The same wavelength of laser light (337 nm) typically employed in MALDI is effective for DIOS. While DIOS is comparable to MALDI with respect to its sensitivity, it has several advantages due to the lack of interfering matrix: low background in the low mass range; uniform deposition of aqueous samples; and simplified sample handling. In addition, the chip-based format can be adapted to automated sample handling, where the laser rapidly scans from spot to spot. DIOS could thus accelerate and simplify high-throughput analysis of low molecular weight compounds, as MALDI has done for macromolecules. Because the masses of many low molecular weight compounds can be measured, DIOS-MS can be applied to the analysis of small molecule transformations, both enzymatic and chemical.

In a number of recent advances with DIOS-MS, the modification of the silicon surface with fluorinated silyating reagents have allowed for ultra-high sensitivity in the yoctomole range (**Figure 2.16**).

2. Ionization

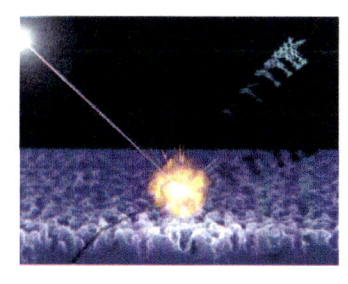

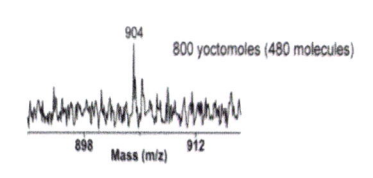

Figure 2.16. Desorption/ionization on nanostructured silicon (DIOS) uses UV laser pulse from a structured silicon surface to generate intact gas phase ions. DIOS allows for small quantities of sample to be analyzed, 800 yoctomoles (480 molecules) of des-arg-bradykinin has been detected.

Fast Atom/Ion Bombardment

Fast atom/ion bombardment, or FAB, is an ionization source (**Figure 2.17**) similar to MALDI in that it uses a matrix and a highly energetic beam of particles to desorb ions from the surface of the matrix. It is important, however, to point out the differences between MALDI and FAB. For MALDI, the energy beam is pulsed laser light, while FAB uses a continuous ion beam. With MALDI, the matrix is typically a solid crystalline, whereas FAB typically has a liquid matrix. It is also important to note that FAB is about 1000 times less sensitive than MALDI.

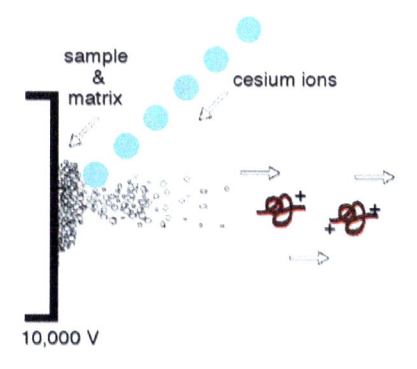

Figure 2.17. Fast atom bombardment (FAB) mass spectrometry, aka liquid secondary ion mass spectrometry (LSIMS).

Fast atom bombardment is a soft ionization source which requires the use of a direct insertion probe for sample introduction, and a beam of Xe neutral atoms or Cs+ ions to sputter the sample and matrix from the direct insertion probe surface. It is common to detect matrix ions in the FAB spectrum as well as the protonated or cationized (i.e. M+Na+) molecular ion of the analyte of interest.

FAB matrix - Facilitating the desorption and ionization process, the FAB matrix is a nonvolatile liquid material that serves to constantly replenish the surface with new sample as it is bombarded by the incident ion beam. By absorbing most of the incident energy, the matrix also minimizes sample degradation from the high-energy particle beam.

Two of the most common matrices used with FAB are m-nitrobenzyl alcohol and glycerol.

The fast atoms or ions impinge on or collide with the matrix causing the matrix and analyte to be desorbed into the gas phase. The sample may already be charged and subsequently transferred into the gas phase by FAB, or it may become charged during FAB desorption through reactions with surrounding molecules or ions. Once in the gas phase, the charged molecules can be propelled electrostatically to the mass analyzer.

Electron Ionization

Electron ionization is one of the most important ionization sources for the routine analysis of small, hydrophobic, thermally stable molecules and is still widely used. Because EI usually generates numerous fragment ions it is a "hard" ionization source. However, the fragmentation information can also be very useful. For example, by employing databases containing over 200,000 electron ionization mass spectra, it is possible to identify an unknown compound in seconds (provided it exists in the database). These databases, combined with current computer

2. Ionization

storage capacity and searching algorithms, allow for rapid comparison with these databases (such as the NIST database), thus greatly facilitating the identification of small molecules.

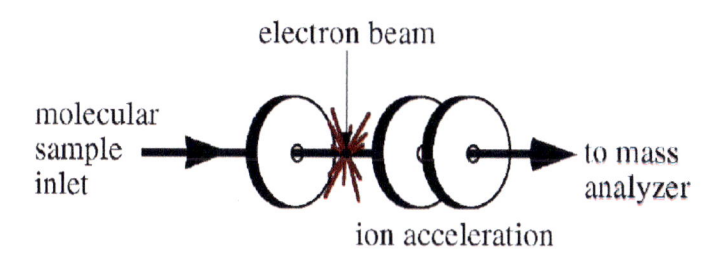

Figure 2.18. Electron ionization (EI) mass spectrometry

The electron ionization source is straightforward in design (**Figure 2.18**). The sample must be delivered as a gas which is accomplished by either "boiling off" the sample from a probe via thermal desorption, or by introduction of a gas through a capillary. The capillary is often the output of a capillary column from gas chromatography instrumentation. In this case, the capillary column provides separation (this is also known as gas chromatography mass spectrometry or GC/MS). Desorption of both solid and liquid samples is facilitated by heat as well as the vacuum of the mass spectrometer. Once in the gas phase the compound passes into an electron ionization source, where electrons excite the molecule, thus causing electron ejection ionization and fragmentation.

The utility of electron ionization decreases significantly for compounds above a molecular weight of 400 Da because the required thermal desorption of the sample often leads to thermal decomposition before vaporization is able to occur. The principal problems associated with thermal desorption in electron ionization are 1) involatility of large molecules, 2) thermal decomposition, and 3) excessive fragmentation.

The method, or mechanism, of electron ejection for positive ion formation proceeds as follows:

- The sample is thermally vaporized.

- Electrons ejected from a heated filament are accelerated through an electric field at 70 V to form a continuous electron beam.

- The sample molecule is passed through the electron beam.

- The electrons, containing 70 V of kinetic energy (70 electron volts or 70 eV), transfer some of their kinetic energy to the molecule. This transfer results in ionization (electron ejection) with the ion internally retaining usually no more than 6 eV excess energy and excess internal energy (5 eV) leads to some degree of fragmentation.

$M + e^- (70\ eV) \rightarrow M^+ (\sim 5\ eV) + 2e^- (\sim 65\ eV)$

$M^+ \rightarrow$ molecular ions + fragment ions + neutral fragments Electron

capture is usually much less efficient than
electron ejection, yet it is sometimes used in the following way for high sensitivity work with compounds having a high electron affinity:

$$M + e^- \rightarrow M^-$$

Chemical Ionization

Chemical Ionization (CI) is applied to samples similar to those analyzed by EI and is primarily used to enhance the abundance of the molecular ion. Chemical ionization uses gas phase ion-molecule reactions within the vacuum of the mass spectrometer to produce ions from the sample molecule. The chemical ionization process is initiated with a reagent gas such as methane, isobutane, or ammonia, which is ionized by electron impact. High gas-pressure in the ionization source results in ion-molecule reactions between the reagent gas ions and reagent gas neutrals. Some of the products of the ion-molecule reactions can react with the analyte molecules to produce ions. A mechanism

2. Ionization

for ionization in CI occurs as follows:

Reagent (R) + e⁻ ⟶ R⁺· + 2e⁻ RH⁺

R⁺ + RH ⟶ + R

RH⁺ + Analyte (A) ⟶ AH⁺ + R

In contrast to EI, an analyte is more likely to provide a molecular ion with reduced fragmentation using CI. However, like EI, samples must be thermally stable since vaporization within the CI source occurs through heating.

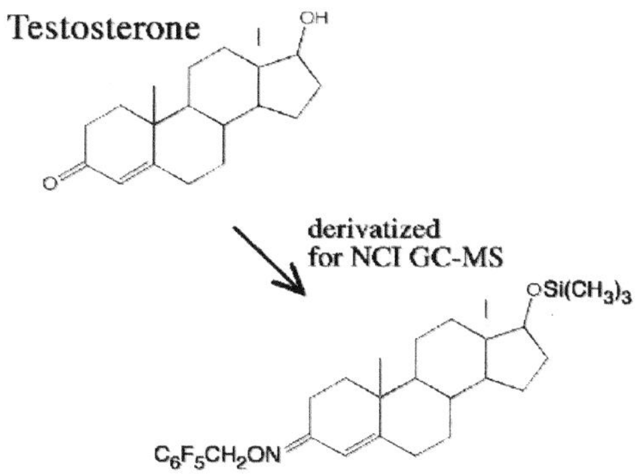

Figure 2.19. The pentafluorobenzyl trimethyl silyl ether derivatives of steroids make them more amenable to high sensitivity measurements using negative chemical ionization.

Negative chemical ionization (NCI) typically requires an analyte that contains electron-capturing moieties (e.g., fluorine atoms or nitrobenzyl groups). Such moieties significantly increase the sensitivity of NCI, in some cases 100 to 1000 times greater than that of electron ionization (EI). NCI is probably one of the most sensitive techniques and is used for a wide variety of small molecules with the caveat that the molecules are often chemically modified with an electron-capturing moiety prior to analysis.

While most compounds will not produce negative ions using EI or CI, many important compounds can produce negative ions and, in some cases, negative EI or CI mass spectrometry is more sensitive and selective than positive ion analysis. Steroids are often modified (**Figure 2.19**) to enhance NCI.

As mentioned, negative ions can be produced by electron capture, and in negative chemical ionization a buffer gas (such as methane) can slow down the electrons in the electron beam allowing them to be captured by the analyte molecules. The buffer gas also stabilizes the excited anions and reduces fragmentation. Therefore, NCI is actually an electron capture process and not what would traditionally be called "chemical ionization".

NanoSIMS

Nano secondary ion mass spectrometry (NanoSIMS) has recently become known for ultrahigh resolution, making subcellular quantitative imaging at the organelle level possible. A remarkable achievement. NanoSIMS (**Figure 2.20**) is an advanced technology that employs a highly focused ion beam to produce high-resolution images with sub-50 nm resolution. This technique has facilitated the visualization of biological samples at the subcellular level, allowing scientists to delve into intricate details.

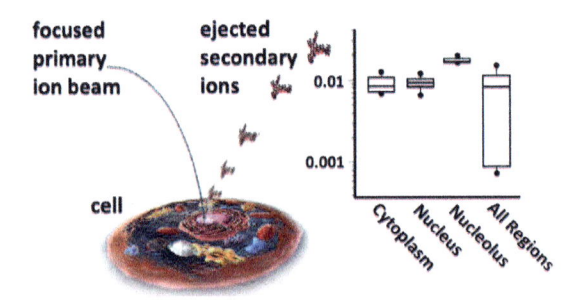

Figure 2.20. NanoSIMS imaging at the sub-50nm level allowing for analysis of organelles and enables absolute quantification of small molecules. (Left) An illustration of the primary ion beam impacting a

2. Ionization

single cell resulting in the ejected secondary ions; (right) box plot data from the cell's respective organelles.

Table 2.5. General Comparison of Ionization Sources.

Ionization Source	Typical Mass Range (Da)	Matrix	Degradation	Complex Mixtures	LC Suitable	Sensitivity
ESI	200,000	none	some ISF	best with LC	excellent	to sub-femtomole
Comments -	Highly compatible for LC/MS; multiple charging useful, soft ionization (some ISF).					
NanoESI	200,000	none	some ISF	best with LC	very low flow rates	to zeptomole
Comments	Very sensitive and requires very low flow rates; applicable to LC/MS; has reasonable salt tolerance (low millimolar); multiple charging useful but significant suppression can occur with mixtures; reasonable tolerance of mixtures; soft ionization (some ISF).					
APCI	1,200	none	thermal	best with LC	excellent	to high attomole
Comments	Excellent LC/MS tool; low salt tolerance (low millimolar); useful for hydrophobic materials.					
APPI	1,200	none	Photo dissociation	best with LC	excellent	to high femtomole
Comments	Excellent LC/MS tool; low salt tolerance (low millimolar); useful for hydrophobic materials.					
MALDI	300,000	Yes, backgrd	photo degradation and matrix reactions	limited	Possible but not common	to high attomole
Comments	Somewhat tolerant of salts; excellent sensitivity; matrix background can be problem for low mass ions; soft ionization (little fragmentation observed); photodegradation possible; suitable for complex mixtures. Limited multiple charging occurs so MS/MS data is not extensive.					
DIOS or NIMS	3,000	None	photo degradation	limited	very limited	to high yoctomole
Comments	Somewhat tolerant of salts; excellent sensitivity; soft ionization (little fragmentation observed); photodegradation possible; suitable for complex mixtures and small molecules.					
FAB	7,000	yes, backgrd	matrix reactions and thermal	limited	very limited	nanomole
Comments	Relatively insensitive; little fragmentation; soft ionization, solubility with matrix required.					
Electron Ionization (EI)	500	none	Thermal	limited unless w/ GC/MS	very limited	picomole
Comments	Good sensitivity; EI fragmentation data generated; NIST database >200,000 compounds available; thermal decomposition an issue thermally unstable molecules.					
Chemical Ionization (CI)	500	none	thermal	limited unless w/ GC/MS	very limited	picomole
Comments	Offers a softer ionization approach over EI yet still requires thermal desorption; negative CI particularly sensitive for perflourinated derivatives; a limited but powerful approach for certain derivatized molecules such as steroids.					
NanoSIMS	<500	none	Primary ion energy transfer	limited	Not suitable	yoctomole
Comments	Relatively harsh ionization yet it offers unprecedented imaging resolution.					

57

Summary

The mass spectrometer can be separated into distinct sections that include the sample inlet, ion source, mass analyzer, and detector. A sample is introduced into the mass spectrometer and is then ionized. The ion source produces ions either by electron ejection, electron capture, cationization, deprotonation or the transfer of a charged molecule from the condensed to the gas phase. MALDI and ESI, especially ESI, have had a profound effect on mass spectrometry because they generate charged intact biomolecules into the gas phase. In comparison to other ionization sources such as APCI, EI, FAB, and CI, the techniques of MALDI and ESI have greatly extended the analysis capabilities of mass spectrometry to a wide range of compounds with detection capabilities ranging from the picomole to the zeptomole level.

Questions and Answers

- What are the primary disadvantages of electron ionization?

  EI requires that the molecules be thermally stable to enable it to be generated in the gas phase. Many molecules, especially of biological origin, degrade when exposed to heat.

- Why is it important to generate the intact ion?

  Generation of intact molecular ions in the gas phase is one of the great achievements of MS. Intact ion generation significantly reduces the ambiguity associated with identifying molecules and is why ESI is so popular.

- Why is ESI useful for studying noncovalent interactions?

  For the same reason it is great at generating intact molecular ions, ESI is a very soft ionization technique.

- What is the purpose of the matrix in MALDI?

  The MALDI matrix serves multiple purposes, it helps screen the analyte molecules from laser photodegradation, it also serves as a proton source for ionization (and with basic matricies, a proton sink). And the matrix (when exposed to pulsed laser radiation) also serves as a medium for intact molecular ejection and ionization in the gas phase.

- What effect does multiple charging have on the mass-to-charge ratio (*m/z*)?

A single charge like a proton, will result in seeing the molecular ion plus a proton in the m/z spectrum. Two protons attaching to the molecule will result in observing (molecular ion plus 2 protons) divided by 2, in the m/z spectrum. And so on…

Useful References

Siuzdak G. The Expanding Role of Mass Spectrometry in Biotechnology 2nd Edition. San Diego: MCC Press, **2006**.

Gross JH. Mass Spectrometry: A Textbook 3rd Edition. Springer Nature, **2017**.

Cole R. (Editor). Electrospray Ionization Mass Spectrometry: Fundamentals, Instrumentation, and Applications. New York: Wiley and Sons, **1997**.

Tanaka K, Waki H, Ido Y, Akita S, Yoshida Y, Yoshida T. Protein and polymer analysis up to m/z 100,000 by laser ionization TOF- MS. Rapid Commun. Mass Spectrom. **1988**, 2, 151.

Karas M & Hillenkamp F. *Laser desorption ionization of proteins with molecular mass exceeding 10,000 Daltons*. Anal. Chem. **1988**, 60, 2299.

Dole M, Mack LL, Hines RL, Mobley RC, Ferguson LD, Alice MB. *Molecular beams of macroions*. J. Chem. Phys. **1968**, 49, 2240.

Fenn JB, Mann M, Meng CK, Wong SF, Whitehouse CM. *Electrospray ionization - principles and practice*. Mass Spectrometry Reviews. **1990**, 9, 37.

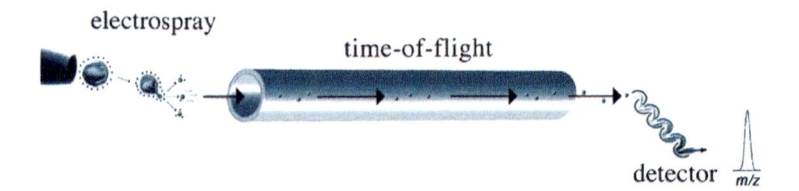

electrospray   time-of-flight

detector m/z

Chapter 3

Mass Analyzers

Ionization to Analysis

With the advent and improvements of atmospheric pressure ionization sources in the 1980s, it became necessary to improve mass analyzer performance with respect to speed, accuracy, and resolution (**Figure 3.1**). More specifically, all mass analyzers have undergone numerous modifications/improvements over the past decade in order to be interfaced with these ionization sources. The biggest challenge came in the interfacing atmospheric pressure ionization sources (760 torr) to analyzers maintained at 10^{-6} to 10^{-11} torr, a remarkable 9 orders of magnitude pressure differential. This chapter will focus on the principles of operation and current performance capabilities of mass analyzers, also touching on detectors and the necessity of low pressures (vacuum) within the mass spectrometer.

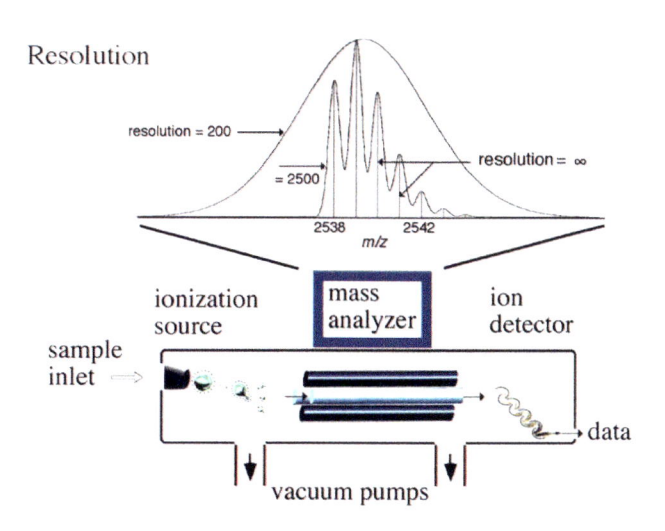

Figure 3.1. Resolution and mass accuracy, the overlaid spectra were calculated for the same molecular formula ($C_{101}H_{145}N_{34}O_{44}$) at resolutions of 200, 2500, and infinity (∞).

Mass Analysis Performance

Analytical instruments in general have variations in their capabilities as a result of their individual design and intended purpose. This is also true for mass spectrometers. While all mass spectrometers rely on a mass analyzer, not all analyzers operate in the same way (**Table 3.1**), some separate ions in space while others separate ions by time. In the most general terms, a mass analyzer measures gas phase ions with respect to their mass-to-charge ratio (*m/z*), where the charge is produced by the addition or loss of a proton(s), cation(s), anion(s) or electron(s). The addition of charge allows the molecule to be affected by electric fields thus allowing its mass measurement. This is an important aspect to remember about mass analyzers -- they measure the *m/z* ratio, not the mass. It is often a point of confusion because if an ion has multiple charges, the *m/z* will be significantly less than the actual mass (**Figures 2.8 and 2.9**). For example, a doubly charged peptide ion of mass 976.5 Daltons (Da) ($C_{37}H_{68}N_{16}O_1^{2+}$) has an *m/z* of 488.3.

Table 3.1. Mass analyzers and ion separation physics.

Mass Analyzers	Event
Quadrupole	scan radio frequency field
Triple Quadrupole	Dual scanning of radio frequency field
Quadrupole Ion Trap	scan radio frequency field
Time-of-Flight (TOF)	time-of-flight correlated directly to ion's *m/z*
Time-of-Flight Reflectron	time-of-flight correlated directly to ion's *m/z*
Quadrupole -TOF	radio frequency field scanning and time-of-flight
Magnetic Sector	magnetic field affects radius of curvature of ions
Fourier Transform MS (FTMS)	translates ion motion to *m/z*
Orbitrap	Electrostatic field motion translated via FTMS to *m/z*

The first mass analyzers, made in the early 1900's, used magnetic fields to separate ions according to their radius of curvature through the magnetic field. The design of modern analyzers has changed significantly in the last five years, now offering much higher accuracy, increased sensitivity, broader mass range, and the ability to provide structural information. Because ionization techniques have evolved, mass analyzers have been forced to change to meet the demands of analyzing a wide range of biomolecular ions with part per million mass accuracy and sub femtomole sensitivity. The characteristics (**Table 3.1**) of these mass analyzers will be covered including their respective performance characteristics including: accuracy, resolution, mass range, tandem analysis capabilities, and scan speed.

Accuracy

Accuracy is the ability with which the analyzer can accurately provide *m/z* information and is largely a function of an instrument's stability and resolution. For example, an instrument with 0.01% accuracy can provide information on a 1000 Da peptide to ±0.1 Da or a 10,000 Da protein to ±1.0 Da. The accuracy varies dramatically from analyzer to analyzer depending on the analyzer type and resolution. An alternative means of describing accuracy is using part per million (ppm) terminology, where 1000 Da peptide to ±0.1 Da could also be described as 1000.00 Da peptide to ± 100 ppm.

Resolution (Resolving Power)

Resolution is the ability of a mass spectrometer to distinguish between ions of different mass-to-charge ratios. Therefore, greater resolution corresponds directly to the increased ability to differentiate ions. The most common definition of resolution is given by the following equation:

$$\text{Resolution} = M/\delta M \qquad \text{Equation 3.1}$$

where M corresponds to *m/z* and δM represents the full width at half maximum (FWHM). An example of resolution measurement is shown in **Figure 3.2** where the peak has an *m/z* of 500 and a FWHM of 1. The resulting resolution is $M/\delta M = 500/1 = 500$.

3. Mass Analyzers

Resolution = M/ΔM

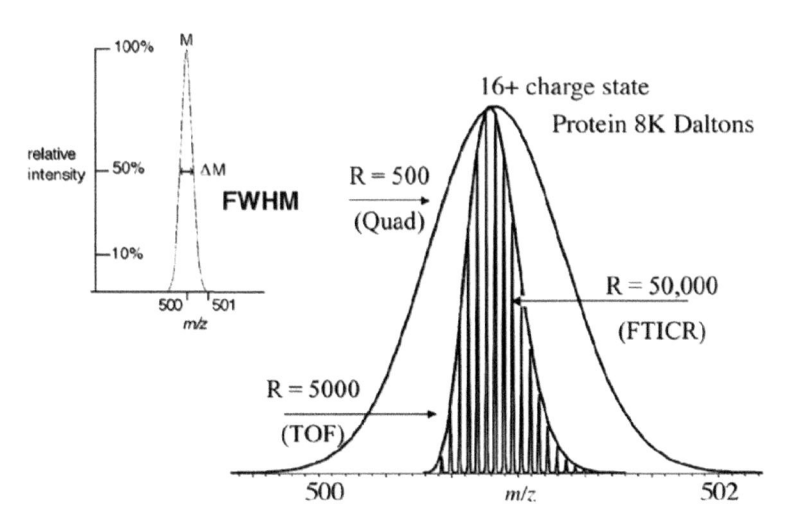

Figure 3.2. The resolution is determined by the measurement of peak's m/z and FWHM, in this case m/z = 500 and the FWHM = 1.

The analyzer's resolving power does, to some extent, determine the accuracy of a particular instrument, as characterized in **Figure 3.2**. The average mass of a molecule is calculated using the weighted average mass of all isotopes of each constituent element of the molecule. The monoisotopic mass is calculated using the mass of the elemental isotope having the greatest abundance for each constituent element. If an instrument cannot resolve the isotopes, it will generate a broad peak with the center representing the average mass. Higher resolution can offer the benefits of separating an ion's individual isotopes or the narrowing of peaks allows a more accurate determination of its position. Resolution however can change depending on the m/z being measured by a particular instrument.

Mass Range

Mass range is the m/z range of the mass analyzer. For instance, quadrupole analyzers typically scan up to m/z 3000. A magnetic sector analyzer typically scans up to m/z 10,000 and time-of-flight analyzers have virtually unlimited m/z range.

Tandem Mass Analysis (MS/MS or MSⁿ)

Tandem mass analysis (**Figure 3.3**) is the ability of the analyzer to separate different molecular ions, generate fragment ions from a selected ion, and then mass measure the fragmented ions. The fragmented ions are used for structural determination of original molecular ions. Typically, tandem MS experiments are performed by colliding a selected ion with inert gas molecules such as argon or helium, and the resulting fragments are mass analyzed. Tandem mass analysis is used to sequence peptides, and structurally characterize carbohydrates, small oligonucleotides, and lipids. And it is widely used in clinical analyses, drug development, and a host of other areas.

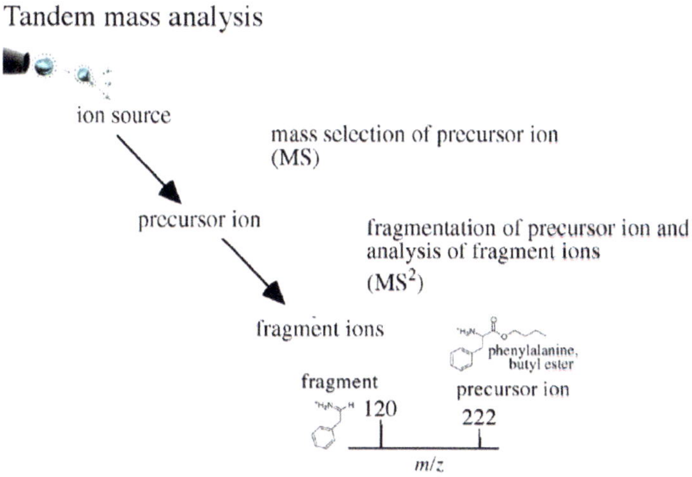

Figure 3.3. Tandem mass spectrometry analysis.

The term "tandem" mass analysis comes from the events occurring either tandem in space or time. Tandem mass analysis in space is performed by consecutive analyzers while tandem mass analysis in time is performed with the same analyzer, which isolates and fragments the ion of interest, then analyzes the fragments. Different analyzers characteristics are summarized in **Table 3.2**.

Scan Speed

Scan speed refers to the rate at which the analyzer scans over a mass range. Some instruments require seconds to perform a full scan, however this can vary depending on the analyzer. Time-of-flight analyzers, for example, complete analyses in milliseconds or less.

Mass Analyzers

It is clear from Chapter 2 that ESI and MALDI are quite different in terms of how ions are generated. ESI creates ions in a continuous stream from charged droplets under atmospheric pressure conditions, for this reason quadrupoles present a well-suited analyzers for ESI since they are tolerant of relatively high pressures (~10^{-5} torr) and are capable of continuously scanning the ESI ion stream. MALDI, on the other hand, generates ions from short, nanosecond laser pulses and is readily compatible with time-of-flight mass analysis, which measures precisely timed ion packets such as those generated from a laser pulse. The most common analyzers are discussed in this section with a description of their respective advantages and disadvantages.

Quadrupoles

Quadrupole mass analyzers (**Figure 3.4**) have been used with EI sources since the 1950's and are still the most common mass analyzers in existence today. Interestingly, quadrupole mass analyzers have found new utility in their capacity to interface with ESI and APCI. Quadrupoles offer three main advantages. They tolerate relatively high pressures. Secondly, quadrupoles have a significant mass range with the capability of analyzing up to an *m/z* of 4000, which is useful because electrospray ionization of proteins and other biomolecules commonly produce charge distributions from *m/z* 1000 to 3500. Finally, quadrupole mass spectrometers are relatively low-cost instruments. Given the complementary features of ESI and quadrupole mass analyzers, it is not surprising that the first successful commercial electrospray instruments were coupled with quadrupole mass analyzers.

Quadrupole mass analyzers are connected in parallel to a radio frequency (RF) generator and a DC potential. At a specific RF field, only ions of a specific *m/z* can pass through the quadrupoles as shown in **Figure 3.4**, where only the ion of *m/z* 100 is detected. In all three cases the DC and RF fields are the same. Therefore, by scanning the RF field a broad *m/z* range (typically 100 to 4000) can be achieved in approximately one second.

Mass Analyzers

Quadrupole Mass Analyzer

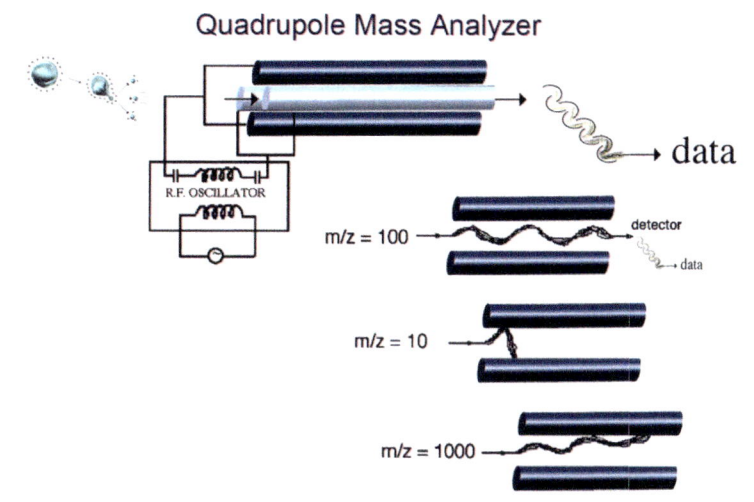

Figure 3.4. Schematic diagram showing arrangement of quadrupole rods and electrical connection to RF generator; a DC potential (not shown) is also superimposed on the rods. A cross-section of a quadrupole mass analyzer taken as it analyzes for m/z 100, 10, and 1000, respectively. It is important to note that both the DC and RF fields are the same in all three cases and only ions with m/z = 100 (top example) traverse the total length of the quadrupole and reach the detector; the other ions are filtered out.

Tandem Mass Spectrometry (MS/MS) with a triple quadrupole

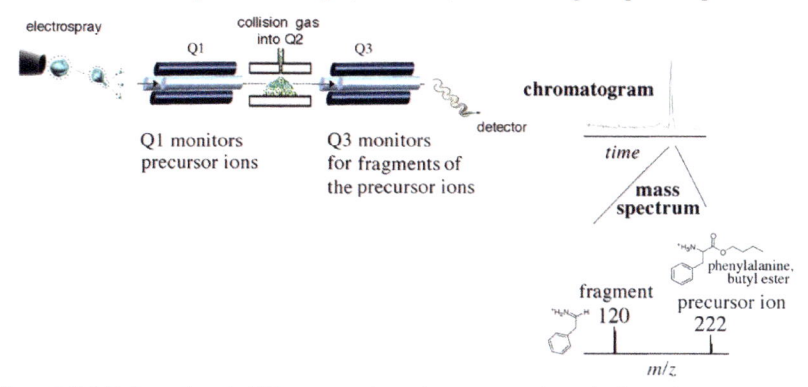

Figure 3.5. A triple quadrupole ESI mass spectrometer possesses ion selection and fragmentation capabilities allowing for tandem mass spectra.

To perform tandem mass analysis with a quadrupole instrument, it is necessary to place three quadrupoles in series. Each quadrupole has a separate function: the first quadrupole (Q1) is used

67

3. Mass Analyzers

to scan across a preset *m/z* range and select an ion of interest. The second quadrupole (Q2), also known as the collision cell, focuses and transmits the ions while introducing a collision gas (argon or helium) into the flight path of the selected ion. The third quadrupole (Q3) serves to analyze the fragment ions generated in the collision cell (Q2) (**Figure 3.5**). A stepwise example of collision-induced dissociation (CID) is shown in **Figure 3.2**.

Quadrupole Ion Trap

The ion trap mass analyzer shown in **Figure 3.6** (roughly the size of a tennis ball) was conceived at the same time as the quadrupole mass analyzer by the same person, Wolfgang Paul. Incidentally, the physics behind both analyzers is similar. However, in an ion trap, rather than passing through a quadrupole analyzer with a superimposed radio frequency field, the ions are trapped in a radio frequency quadrupole field. One method of using an ion trap for mass spectrometry involves generating ions internally with EI, followed by mass analysis. Another, more popular, method of using an ion trap for mass spectrometry involves generating ions externally with ESI or MALDI and using ion optics for sample injection into the trapping volume. The quadrupole ion trap typically consists of a ring electrode and two hyperbolic endcap electrodes (**Figure 3.6**). The motion of the ions induced by the electric field on these electrodes allows ions to be trapped or ejected from the ion trap. In the normal mode, the radio frequency is scanned to resonantly excite and therefore eject ions through small holes in the endcap to a detector. As the RF is scanned to higher frequencies, higher *m/z* ions are excited, ejected, and detected.

A particularly useful feature of ion traps is that it is possible to isolate one ion species by ejecting all others from the trap. The isolated ions can subsequently be fragmented by collisional activation and the fragments detected. The primary advantage of quadrupole ion traps is that multiple collision induced dissociation experiments can be performed quickly without having multiple analyzers, such that real time LC-MS/MS is now routine. Other important advantages of quadrupole ion traps include their compact size, and their ability to trap and accumulate ions to provide a better ion signal.

Mass Analyzers

Quadrupole ion traps have been utilized in several applications ranging from electrospray ionization MS^n (**Figure 3.6**) of biomolecules to their more recent interface with MALDI. MS^n allows for multiple MS/MS experiments to be performed on subsequent fragment ions, providing additional fragmentation information. Yet, ion traps most important application has been in the characterization of proteins. LC-MS/MS experiments are performed on proteolytic digests which provide both MS and MS/MS information. This information allows for protein identification and post-translational modification characterization. The mass range (~4000 m/z) of commercial LC-traps is well matched to m/z values generated from the electrospray ionization of peptides and the resolution allows for charge state identification of multiply-charged peptide ions. Quadrupole ion trap mass spectrometers can analyze peptides from a tryptic digest present at the 20-100 fmol level. Another asset of the ion trap technique for peptide analysis is the ability to perform multiple stages of mass spectrometry, which can significantly increase the amount of structural information.

Tandem Mass Spectrometry (MS/MS) with an ion trap

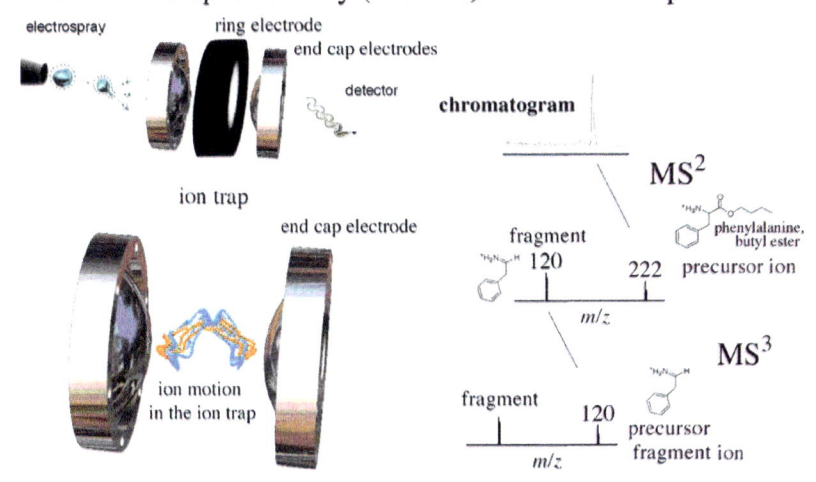

Figure 3.6. Ions inside a 3D ion trap mass analyzer can be analyzed to produce a mass spectrum, or a particular ion can be trapped inside and made to undergo collisions to produce fragmentation information.

Linear Ion Trap

A linear ion trap differs from the 3D ion trap (**Figure 3.7**) as it confines ions along the axis of a quadrupole mass analyzer using a two-dimensional (2D) radio frequency (RF) field with potentials applied to end electrodes. The primary advantage of a linear trap over the 3D trap is the larger analyzer volume lends itself to a greater dynamic range and an improved range of quantitative analysis.

Quadrupole Linear Ion Trap Mass Analyzer

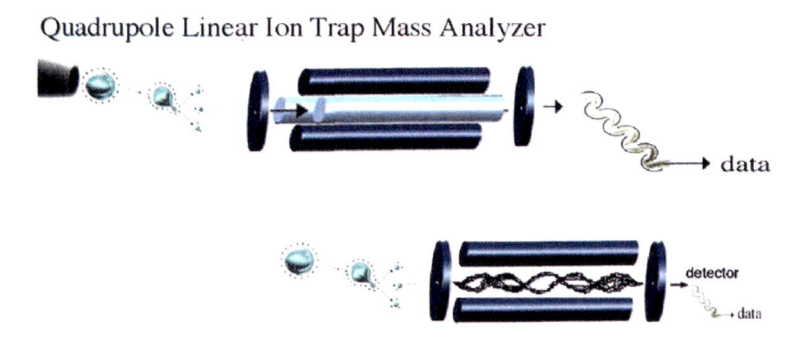

Figure 3.7. A linear ion trap mass analyzer confines the ions along the axis of quadrupoles using a 2D radio frequency and stopping potentials on the end electrodes.

Given the power of the ion trap the major limitations of this device that keep it from being the ultimate tool for pharmacokinetics and proteomics include the following: 1) the ability to perform high sensitivity triple quadrupole-type precursor ion scanning and neutral loss scanning experiments is not possible with ion traps. 2) The upper limit on the ratio between precursor m/z and the lowest trapped fragment ion is ~0.3 (also known as the "one third rule"). An example of the one third rule is that fragment ions of m/z 900 will not be detected below m/z 300, presenting a significant limitation for *de novo* sequencing of peptides. 3) The dynamic range of ion traps are limited because when too many ions are in the trap, space charge effects diminish the performance of the ion trap analyzer. To get around this, automated scans can rapidly count ions before they go into the trap, therefore limiting the number of ions getting in. Yet this approach presents a problem when an ion of interest is accompanied by a large background ion population.

Double-Focusing Magnetic Sector

The earliest mass analyzers separated ions with a magnetic field. In magnetic analysis, the ions are accelerated into a magnetic field using an electric field. A charged particle traveling through a magnetic field will travel in a circular motion with a radius that depends on the speed of the ion, the magnetic field strength, and the ion's *m/z*. A mass spectrum is obtained by scanning the magnetic field and monitoring ions as they strike a fixed-point detector. A major limitation of magnetic analyzers is their relatively low resolution. In order to improve this, magnetic instruments were modified with the addition of an electrostatic analyzer to focus the ions. These are called double-sector or two-sector instruments. The electric sector serves as a kinetic energy focusing element allowing only ions of a particular kinetic energy to pass through its field irrespective of their mass-to-charge ratio. Thus, the addition of an electric sector allows only ions of uniform kinetic energy to reach the detector, thereby decreasing the kinetic energy spread, which in turn increases resolution. It should be noted that the corresponding increase in resolution does have its costs in terms of sensitivity. These double-focusing (**Figure 3.8**) mass analyzers are used with ESI, FAB and EI ionization, however they are not widely used today primarily due to their large size and the success of time-of-flight, quadrupole and FTMS analyzers with ESI and MALDI.

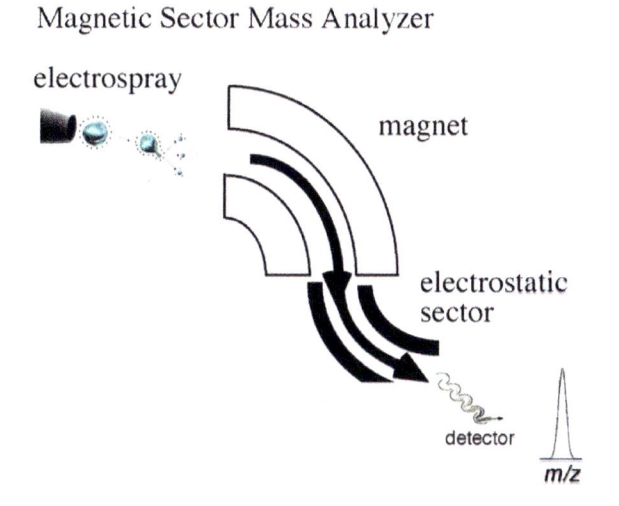

Figure 3.8. A two-sector double-focusing instrument.

Time-of-Flight Tandem MS

The linear time-of-flight (TOF) mass analyzer (**Figure 3.9**) is the simplest mass analyzer. It has enjoyed a renaissance with the invention of MALDI and its recent application to electrospray and even gas chromatography electron ionization mass spectrometry (GC/MS). Time-of-flight analysis is based on accelerating a group of ions to a detector where all of the ions are given the same amount of energy through an accelerating potential. Because the ions have the same energy, but a different mass, the lighter ions reach the detector first because of their greater velocity, while the heavier ions take longer due to their heavier masses and lower velocity. Hence, the analyzer is called time-of-flight because the mass is determined from the ions' time of arrival. Mass, charge, and kinetic energy of the ion all play a part in the arrival time at the detector. Since the kinetic energy (KE) of the ion is equal to $1/2\, mv^2$, the ion's velocity can be represented as $v = d/t = (2KE/m)^{1/2}$. The ions will travel a given distance d, within a time t, where t is dependent upon the mass-to-charge ratio (m/z). In this equation, $v = d/t = (2KE/m)^{1/2}$, assuming that $z = 1$. Another representation of this equation to more clearly present how mass is determined is $m = 2t^2\, KE/d^2$ where KE is constant.

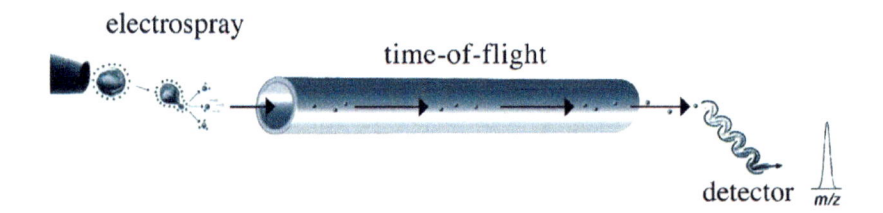

Figure 3.9. Time-of-flight and time-of-flight reflectron mass analyzers. The TOF analyzer has virtually unlimited mass range, while the TOF reflectron has mass range up to m/z ~10,000. It should be noted that most detectors have a limited mass range.

The time-of-flight (TOF) reflectron (**Figure 3.10**) is now widely used for ESI, MALDI, and more recently for electron ionization in GC/MS applications. It combines time-of-flight technology with an electrostatic mirror. The reflectron serves to increase the amount of time (t) ions need to reach the detector while reducing their kinetic energy distribution, thereby reducing the temporal distribution Δt. Since resolution is defined by the mass of a peak divided by the width

Mass Analyzers

of a peak or m/Δm (or t/Δt since m is related to t), increasing t and decreasing Δt results in higher resolution. Therefore, the TOF reflectron offers high resolution over a simple TOF instrument by increasing the path length and kinetic energy focusing through the reflectron. It should be noted that the increased resolution (typically above 5000) and sensitivity on a TOF reflectron does decrease significantly at higher masses (typically above 5000 *m/z*).

Another type of tandem mass analysis, MS/MS, is also possible with MALDI TOF reflectron mass analyzers. MS/MS is accomplished by taking advantage of MALDI fragmentation that occurs following ionization, or post-source decay (PSD). Time-of-flight instruments alone will not separate post-ionization fragment ions from the same precursor ion because both the precursor and fragment ions have the same velocity and thus reach the detector at the same time. The reflectron takes advantage of the fact that the fragment ions have different kinetic energies and separates them based on how deeply the ions penetrate the reflectron field, thus producing a fragment ion spectrum (**Figure 3.10** and **3.11**).

Time-of-Flight Reflectron Mass Analyzer

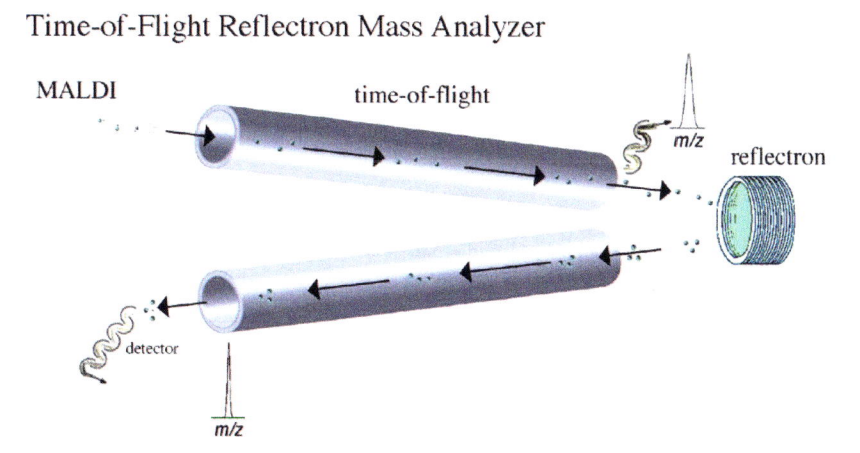

Figure 3.10. A MALDI time-of-flight reflectron mass analyzer and its ability to improve resolution over time-of-flight analysis with the reflectron. The TOF reflectron mass analyzer with an ESI ion source has also gained wide use due to the fast acquisition rates (milliseconds), good mass range (up to ~10,000 *m/z*) and accuracy on the order of 5 part per million (ppm).

3. Mass Analyzers

It is worth noting that ESI is readily adapted to TOF reflectron analyzers, where the ions from the continuous ESI source can be stored in the hexapole (or octapole) ion guide then pulsed into the TOF analyzer. Thus, the necessary electrostatic pulsing creates a time zero from which the TOF measurements can begin.

Time-of-Flight Reflectron Mass Analyzer

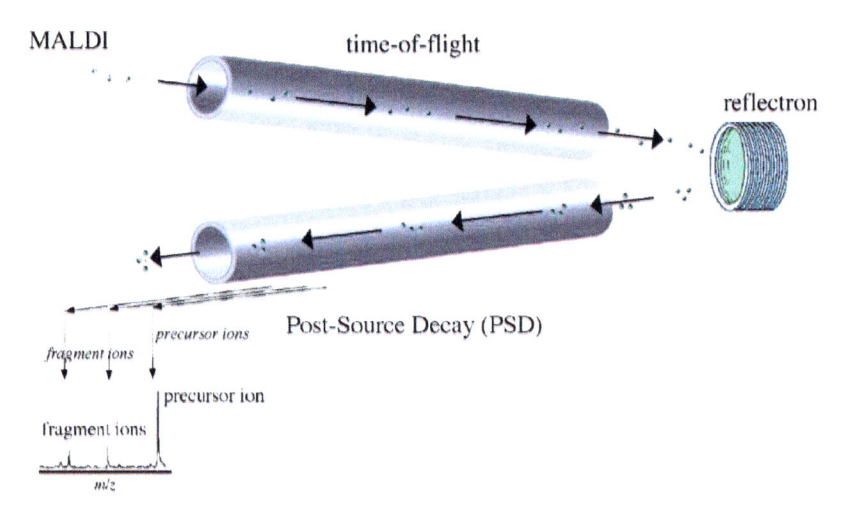

Figure 3.11. A MALDI time-of-flight reflectron mass analyzer and its ability to generate fragmentation information. Fragmentation analysis from a MALDI TOF reflectron is known as post-source decay or PSD.

MALDI with Time-of-Flight Analysis

In the initial stages of MALDI–TOF development, these instruments had relatively poor resolution which severely limited their accuracy. An innovation that has had a dramatic effect on increasing the resolving power of MALDI time-of-flight instruments has been delayed extraction (DE), as shown in **Figure 3.12**. In theory, delayed extraction is a relatively simple means of cooling and focusing the ions immediately after the MALDI ionization event, yet in practice it was initially a challenge to pulse 10,000 volts on and off within a nanosecond time scale. In traditional MALDI instruments, the ions were accelerated out of the ionization source immediately as they were formed. However, with delayed extraction the ions are allowed to "cool" for ~150 nanoseconds before being accelerated to the analyzer. This cooling period generates a set of ions with a much smaller kinetic energy distribution, ultimately reducing the temporal spread of ions once they enter the TOF analyzer. Overall, this results in increased resolution and accuracy. The benefits of delayed extraction significantly diminish with larger macromolecules such as proteins (>30,000 Da).

Delayed Extraction with MALDI

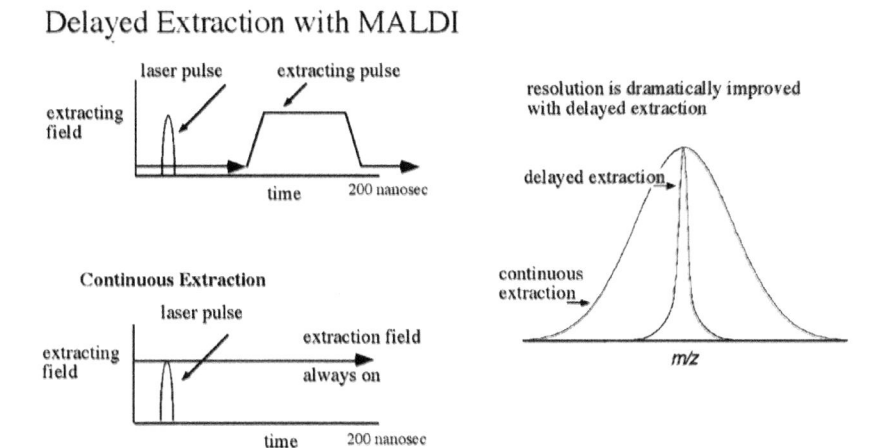

Figure 3.12. Delayed extraction (DE) is a technique applied in MALDI which allows ions to be extracted from the ionization source after a cooling period of ~150 nanoseconds. This cooling period effectively narrows the kinetic energy distribution of the ions, thus providing higher resolution than in continuous extraction techniques.

Quadrupole Time-of-Flight MS

Quadrupole-TOF mass analyzers are typically coupled to electrospray ionization sources and more recently they have been successfully coupled to MALDI. The ESI quad-TOF (**Figure 3.13**) combines the stability of a quadrupole analyzer with the high efficiency, sensitivity, and accuracy of a time-of-flight reflectron mass analyzer. The quadrupole can act as any simple quadrupole analyzer to scan across a specified *m/z* range. However, it can also be used to selectively isolate a precursor ion and direct that ion into the collision cell. The resultant fragment ions are then analyzed by the TOF reflectron mass analyzer. Quadrupole-TOF exploits the quadrupole's ability to select a particular ion and the ability of TOF-MS to achieve simultaneous and accurate measurements of ions across the full mass range. This is in contrast to conventional analyzers, such as tandem quadrupoles, which must scan over one mass at a time. Quadrupole-TOF analyzers offer significantly higher sensitivity and accuracy over tandem quadrupole instruments when acquiring full fragment mass spectra.

The quadrupole-TOF instrument can use either the quadrupole or TOF analyzers independently or together for tandem MS experiments. The TOF component of the instrument has an upper m/z limit in excess of 10,000. The high resolving power (~10,000) of the TOF also enables good mass measurement accuracy on the 10 ppm level. Due to its high accuracy and sensitivity, the ESI quad-TOF mass spectrometer is being incorporated into both proteomics and pharmacokinetics problem solving.

Quadrupole Time-of-Flight Mass Analyzer

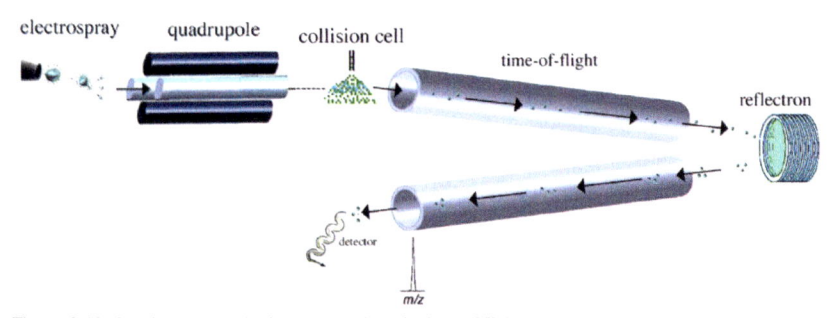

Figure 3.13. An electrospray ionization quadrupole time-of-flight mass spectrometer.

Fourier Transform Ion Cyclotron Resonance MS (FTMS)

FTMS is based on the principle of monitoring a charged particle's orbiting motion in a magnetic field (**Figure 3.14-15**). While the ions are orbiting, a pulsed radio frequency (RF) signal is used to excite them. This RF excitation allows the ions to produce a detectable image current by bringing them into coherent motion and enlarging the radius of the orbit. The image current generated by all

Mass Analyzers

of the ions can then be Fourier-transformed to obtain the component frequencies of the different ions, which correspond to their *m/z*. Because the frequencies can be obtained with high accuracy, their corresponding *m/z* can also be calculated with high accuracy. It is important to note that a signal is generated only by the coherent motion of ions under ultra-high vacuum conditions ($10^{-11} - 10^{-9}$ Torr). This signal must be measured for a minimum amount of time (typically 500 ms to 1 second) to provide high resolution. As pressure increases, signal decays faster due to loss of coherent motion due to collisions (e.g. in ~ <150 ms) and does not allow for high resolution measurements (**Figure 3.14**).

Fourier transform ion cyclotron resonance mass analyzer

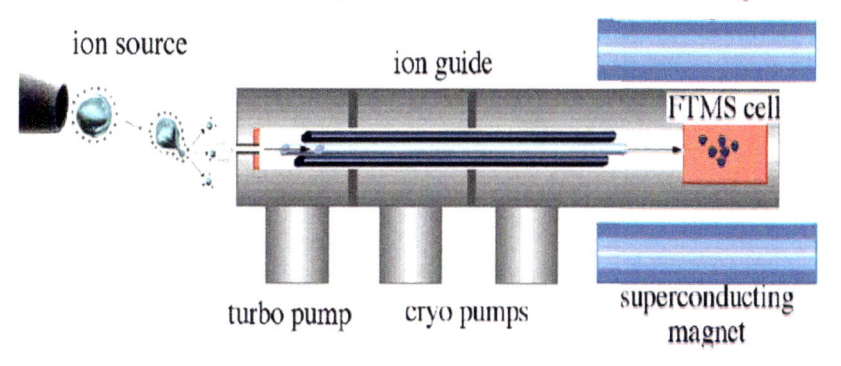

Figure 3.14. A side view of an FTMS instrument with ESI source. The ESI ions are formed and guided into the analyzer cell using a single stage quadrupole rod assembly. The analyzer cell rests in the superconducting magnet (diagram courtesy IonSpec Corporation).

Ions undergoing coherent cyclotron motion between two electrodes are illustrated in **Figure 3.14**. As the positively charged ions move away from the top electrode and closer to the bottom electrode, the electric field of the ions induces electrons in the external circuit to flow and accumulate on the bottom electrode. On the other half of the cyclotron orbit, the electrons leave the bottom electrode and accumulate on the top electrode as the ions approach. The oscillating flow of electrons in the external circuit is called an image current. When a mixture of ions with different *m/z* values are all simultaneously accelerated, the image current signal at the output of the amplifier is a composite transient signal with frequency components representing each m/z value. In short, all of the ions trapped in the analyzer cell are excited into a higher cyclotron orbit, using a radio frequency pulse. The composite transient image current signal of the ions as they relax is acquired by a computer and a Fourier transform is used to separate out the individual cyclotron frequencies. The effect of pressure on the signal and resolution is demonstrated in **Figure 3.16**.

Ion Cyclotron Motion with Fourier transformation

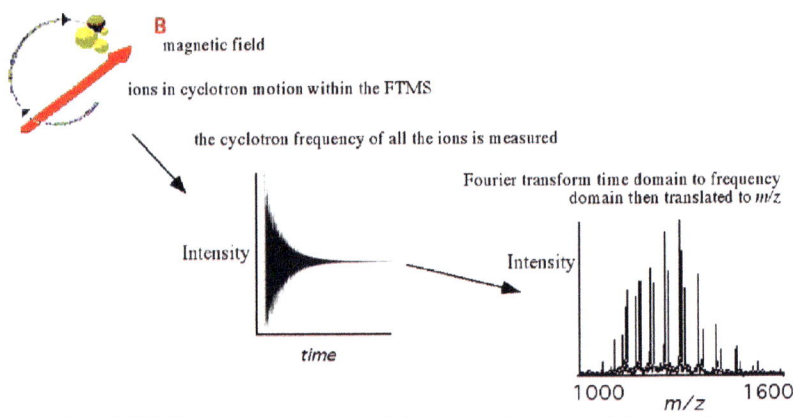

Figure 3.15. ESI FTMS data generated on multiple proteins, the sinusoidal composite image current for all m/z ions can be Fourier transformed to measure frequencies (and m/z values).

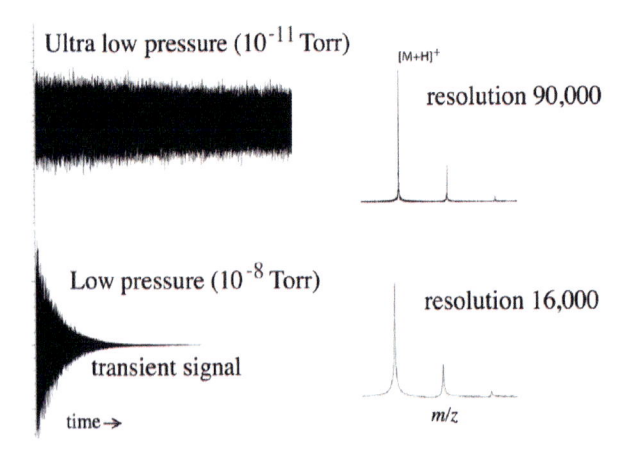

Figure 3.16. Pressure effect on transient signal and resolution.

In addition to high resolution, FTMS also offers the ability to perform multiple collision experiments (MSⁿ). FTMS is capable of ejecting all but the ion of interest. The selected ion is then subjected to a collision gas (or another form of excitation such as laser light or electron capture) to induce fragmentation. Mass analysis can then be carried out on the fragments to generate a fragmentation spectrum. The high resolution of FTMS/MS also yields high-accuracy fragment masses.

Protein isotope pattern from an FTMS

m/z spectrum

isotope spacing (0.2) determines charge state z, which = 1/spacing = 1/0.2 = 5 = z

$(m/z \times z)$ - # protons = protein mass
$(2004 \times 5) - 5 = 10,015$ Da

deconvolution

10,015 Da

molecular weight spectrum

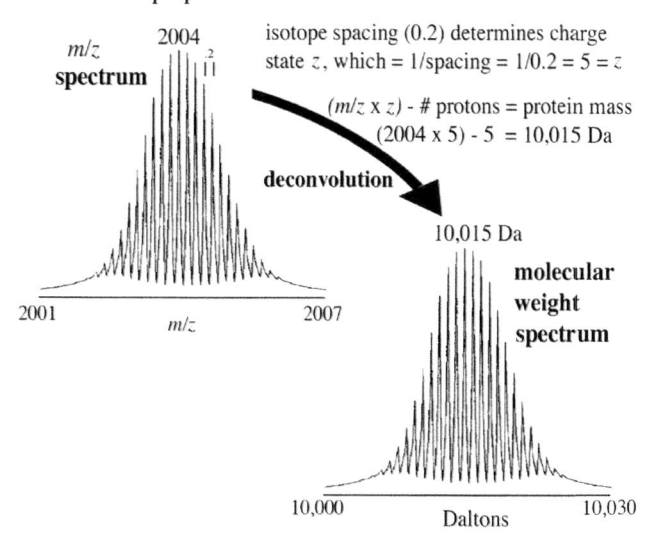

Figure 3.17. A demonstration of deconvolution from an FTMS mass spectrum of a 10 KDa protein at a resolution of 30,000. The cluster of peaks represents the isotope distribution of a protein and the 0.2 *m/z* isotope spacing indicates a 5+ charge state.

FTMS is a relative neophyte to biomolecular analysis, yet many of its advantages have become a mainstay of the research community, especially the orbitrap mass analyzer. It is now becoming more common to couple ultrahigh resolution (>10^5) FTMS to a wide variety of ionization sources, including MALDI, ESI, APCI, and EI. The result of an FTMS analyzer's high resolving power is high accuracy (often at the part per million level) as illustrated for a protein in **Figure 3.17** where individual isotopes can be observed. The Fourier transform of the ICR signal greatly enhances the utility of ICR by simultaneously measuring all the overlaying frequencies produced by the ions within the ICR cell. The individual frequencies can then be easily and accurately translated into the ion's *m/z*.

In general, increasing magnetic field (B) has a favorable effect on performance. The Fourier transform of the ICR signal, by measuring overlaying frequencies simultaneously, allows for high resolution and high mass accuracy without compromising sensitivity. This is in sharp contrast to double sector instruments that suffer from a loss in sensitivity at the highest resolution and accuracy. The high-resolution capabilities of FTMS are directly related to the magnetic field of the FTMS superconducting magnet, with the resolution increasing as a linear function of the field. The ion capacity as well as MS/MS kinetic energy experiments increases as a square of the magnetic therefore improving dynamic range and fragmentation data. One challenge in increasing B is the magnetic mirror effect where ion transmission to the inside of magnetic field becomes more difficult due to magnetic field lines. Also, manufacturing high field magnets with larger bores and excellent field homogeneity (in the ICR housing) becomes technically more difficult.

FTMS instrumentation is affected by the magnetic field in the following ways:

FTMS attribute	Effect of Magnetic Field Strength (B)	What it means:
Resolution ($m/\Delta m$)	Directly proportional to **B**	Improves mass accuracy and the ability to get isotopic resolution on large macromolecules
Kinetic Energy	Directly proportional to **B^2**	Increases the fragmentation and also ability to fragment larger macromolecules
Ion capacity	Directly proportional to **B^2**	Can store more ions before space-charge adversely affects performance

Mass Analyzers

Ion cyclotron resonance

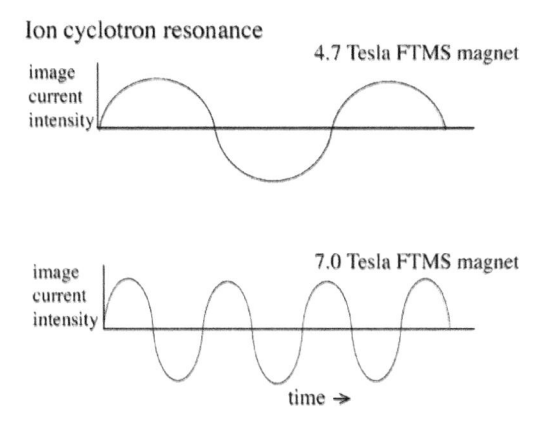

[Figure showing image current intensity vs time for 4.7 Tesla FTMS magnet (lower frequency sine wave) and 7.0 Tesla FTMS magnet (higher frequency sine wave)]

Figure 3.18. Fourier transform ion cyclotron frequency increases with magnetic field strength. The increased frequency improves accuracy as it allows for more measurements to average.

Since ion frequency = $K*B*z/m$, larger magnetic fields provide a higher frequency for the same m/z, therefore more data points are generated to define the frequency more precisely which ultimately increases accuracy (**Figure 3.18**).

Quadrupole and Quadrupole Linear Ion Trap FTMS

Quadrupole-FTMS and quadrupole linear ion trap-FTMS mass analyzers that have recently been introduced are typically coupled to electrospray ionization sources. The quad-FTMS combines the stability of a quadrupole analyzer with the high accuracy of a FTMS. The quadrupole can function as a simple quadrupole analyzer to scan across a specified m/z range. However, it can also be used to selectively isolate a precursor ion and direct that ion into the collision cell or the FTMS. The resultant precursor and fragment ions can then be analyzed by the FTMS.

Performing MS/MS experiments outside the magnet presents some advantages since high resolution in FTMS is dependent on the presence of high vacuum. MS/MS experiments involve collisions at a transiently high pressure ($10^{-6} - 10^{-7}$ Torr) that then has to be reduced to achieve high resolution ($10^{-10} - 10^{-9}$ Torr). Performing MS/MS experiments outside the cell is thus faster since the ICR cell can be

Orbitrap FTMS

The orbitrap is one of the newest mass analyzers, having been introduced in the early 2000s. An Orbitrap (**Figure 3.19**) is a high-resolution FTMS that is now commonly used in metabolomics, proteomics, and other fields. The instrument operates by using an electric field to trap ions inside a cylindrical ion trap, known as the Orbitrap. The Orbitrap consists of a central spindle electrode and a surrounding outer electrode that creates a harmonic oscillation of the ions in the radial direction. The frequency of the oscillation is dependent on the mass-to-charge ratio of the ions, allowing for the determination of the mass-to-charge ratio of the ions with high accuracy and resolution.

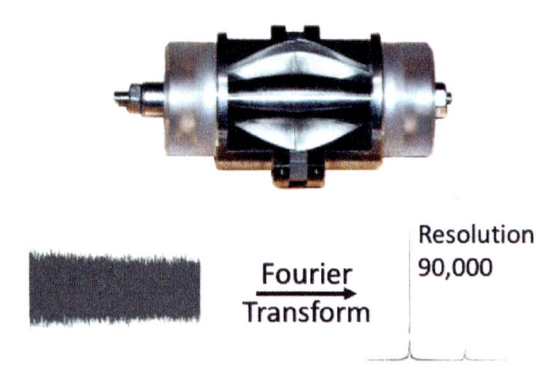

Figure 3.19. Fourier transform orbitrap mass spectrometer was introduced in the 2000s and has become a popular high resolution mass spectrometer.

To perform a measurement, a sample is ionized, and the resulting ions are injected into the Orbitrap. The ions are then trapped and allowed to oscillate in the electric field. As the ions oscillate, their image current induces a current in the outer electrode that is measured by a detector. The resulting signal is then Fourier transformed to obtain the mass spectrum of the sample. The Orbitrap is capable of measuring mass-to-charge ratios with a resolution of up to 500,000, making it one of the higher resolution mass analyzers.

Table 3.2. A general comparison of mass analyzers typically used for ESI.

	Quadrupole	Orbitrap	TOF Reflectron	FTMS	Q-TOF
Accuracy	0.01% (100 ppm)	<0.0005% (<5 ppm)	<0.0005% (<5 ppm)	<0.0005% (<5 ppm)	<0.0005% (<5 ppm)
Resolution	4000	100,000	30,000	Up to 1M	30,000
m/z Range	4000	10,000	10,000	10,000	10,000
Scan Speed	Seconds	milliseconds	milliseconds	seconds	seconds
Tandem MS	MS^2 (triple quad)	MS^n	MS^2	MS^n	MS^2
Tandem MS Comments	Good accuracy Good resolution Low-energy collisions	Excellent accuracy & resolution Low-energy collisions	Precursor ion selection is limited to a wide mass range	Excellent accuracy & resolution of product ions	Excellent accuracy Good resolution Low-energy collisions High sensitivity
General Comments	Low cost Ease of switching between pos/neg ions	High accuracy and resolution tandem MS	Good accuracy Good resolution	High resolution Tandem MS Requires high vacuum & super-conducting magnet	Known for high sensitivity & accuracy when used for tandem MS measurements

*Many of these values vary with respect to manufactured instrument.

Detectors

Once the ions are separated by the mass analyzer, they reach the ion detector (**Figures 3.1** and **3.20-23**), which generates a current signal from the incident ions. The most used detector is the electron multiplier, which transfers the kinetic energy of incident ions to a surface that in turn generates secondary electrons. However, a variety of approaches (**Table 3.3**) are used to detect ions depending on the type of mass spectrometer.

Electron Multiplier

Perhaps the most common means of detecting ions involves an electron multiplier (**Figure 3.20**), which is made up of a series (12 to 24) of aluminum oxide (Al_2O_3) dynodes maintained at ever increasing potentials. Ions strike the first dynode surface causing an emission of electrons. These electrons are then attracted to the next dynode held at a higher potential and therefore more secondary electrons are generated. Ultimately, as numerous dynodes are involved, a cascade of electrons is formed that results in an overall current gain on the order of one million or higher.

Electron Multiplier

one ion in

A series of dynodes at increasing potentials produce a cascade of electrons.

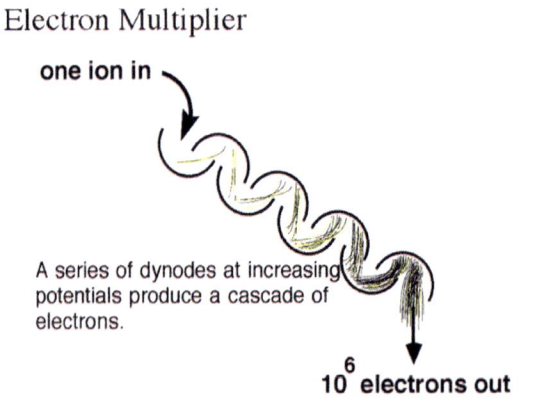

10^6 electrons out

Figure 3.20. Diagrammatic representation of an electron multiplier and the cascade of electrons that results in a 106 amplification of current in a mass spectrometer.

The high energy dynode (HED) uses an accelerating electrostatic field to increase the velocity of the ions. Since the signal on an electron multiplier is highly dependent on ion velocity, the HED serves to increase signal intensity and therefore sensitivity.

Faraday Cup

A Faraday cup (**Figure 3.21**) involves an ion striking the dynode (BeO, GaP, or CsSb) surface which causes secondary electrons to be ejected. This temporary electron emission induces a positive charge on the detector and therefore a current of electrons flowing toward the detector. This detector is not particularly sensitive, offering limited amplification of signal, yet it is tolerant of relatively high pressure.

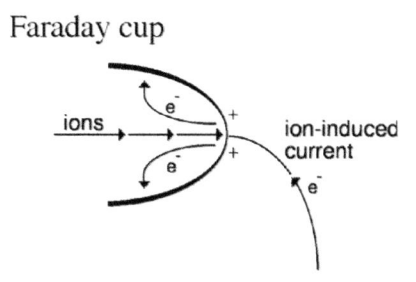

Figure 3.21. Faraday cups convert the striking ion into a current by temporarily emitting electrons creating a positive charge and simultaneously adsorbing the striking ion.

Photomultiplier Conversion Dynode

The photomultiplier conversion dynode detector (**Figure 3.22**) is not as commonly used as the electron multiplier, yet it is similar in design where the secondary electrons strike a phosphorus screen instead of a dynode. The phosphorus screen releases photons which are detected by the photomultiplier. Photomultipliers also operate like the electron multiplier where the striking of the photon on a scintillating surface result in the release of electrons that are then amplified using the cascading principle. One advantage of the conversion dynode is that the photomultiplier tube is sealed in a vacuum, unexposed to the environment of the mass spectrometer and thus the possibility of contamination is removed. This improves the lifetimes of these detectors over electron multipliers. A five-year or greater lifetime is typical, and they have a similar sensitivity to the electron multiplier.

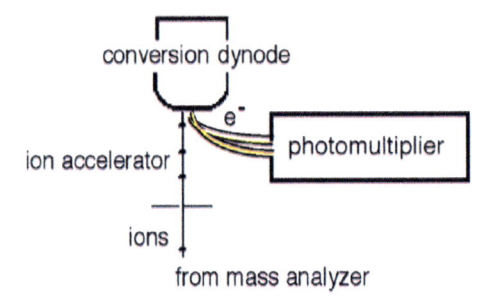

Figure 3.22. Scintillation counting with a conversion dynode and a photomultiplier relies on the conversion of the ion (or electron) signal into light. Once the photon(s) are formed, detection is performed with a photomultiplier.

Array Detector

An array detector is a group of individual detectors aligned in an array format. The array detector, which spatially detects ions according to their different m/z, has been typically used on magnetic sector mass analyzers. Spatially differentiated ions can be detected simultaneously by an array detector. The primary advantage of this approach is that, over a small mass range, scanning is not necessary and therefore sensitivity is improved.

Charge (or Inductive) Detector

Charge detectors simply recognize a moving charged particle (an ion) through the induction of a current on the plate as the ion moves past. A typical signal is shown in **Figure 3.23**. This type of detection is widely used in FTMS to generate an image current of an ion. Detection is independent of ion size and therefore has been used on particles such as whole viruses.

Charge Detection

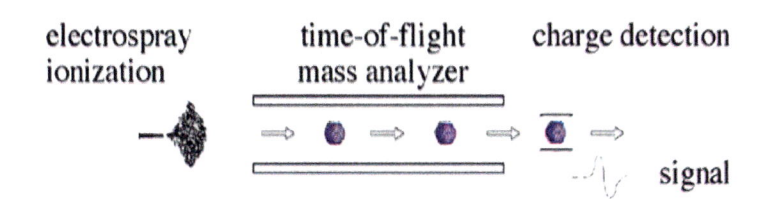

Figure 3.23. Illustration of the operation of a mass spectrometer with a charge detector; as a charged species passes through a plate it induces a current on the plate.

Table 3.3. General comparison of detectors.

Detector	Advantages	Disadvantages
Faraday Cup	Good for checking ion transmission and low sensitivity measurements	Low amplification (≈ 10)
Photomultiplier Conversion Dynode (Scintillation Counting)	Robust; Long lifetime (>5 years); Sensitive ($\approx$ gains of 10^6)	Cannot be exposed to light while in operation
Electron Multiplier	Robust; Fast response; Sensitive ($\approx$ gains of 10^6)	Shorter lifetime than scintillation counting (~3 years)
High Energy Dynodes with electron multiplier	Increases high mass sensitivity	May shorten lifetime of electron multiplier
Array	Fast and sensitive	Reduces resolution; Expensive
Charge Detection	Detects ions independent of mass and velocity	Limited compatibility with most existing instruments

Vacuum in the Mass Spectrometer

All mass spectrometers need a vacuum to allow ions to reach the detector without colliding with other gaseous molecules or atoms. If such collisions did occur, the instrument would suffer from reduced resolution and sensitivity. Higher pressures may also cause high voltages to discharge to ground which can damage the instrument, its electronics, and/or the computer system running the mass spectrometer. An extreme leak, basically an implosion, can seriously damage a mass spectrometer by destroying electrostatic lenses, coating the optics with pump oil, and damaging the detector. In general, maintaining a good vacuum is crucial to obtaining high quality spectra.

Vacuum systems

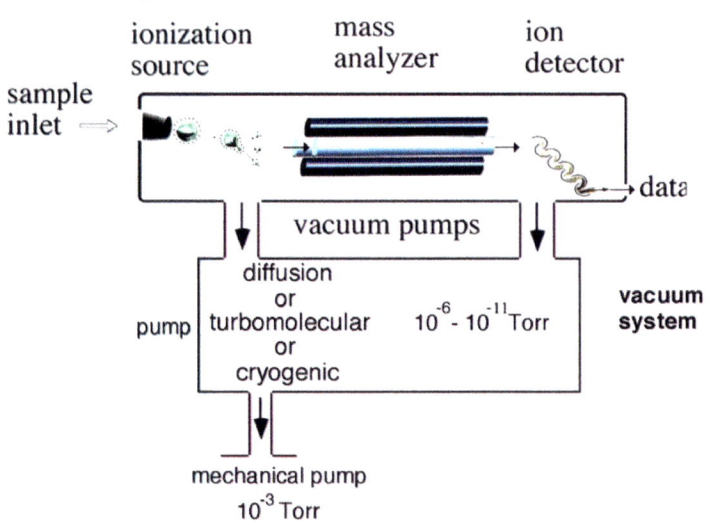

Figure 3.24. A well-maintained vacuum is essential to the function of a mass spectrometer. A couple of the different types of vacuum systems are illustrated.

One of the first obstacles faced by the originators of mass spectrometry was coupling the sample source to a mass spectrometer. The sample is initially at atmospheric pressure (760 torr) before being transferred into the mass spectrometer's vacuum (~10^{-6} torr), which represents approximately a billion-fold difference in pressure. One approach is to introduce the sample through a capillary column (GC) or through a small orifice directly into the instrument. Another approach is to evacuate the sample chamber through a vacuum lock (MALDI) and once a reasonable vacuum is achieved (< 10^{-2} torr) the sample can be presented to the primary vacuum chamber (< 10^{-5} torr).

A mass spectrometer is shown in **Figure 3.24** with three alternative pumping systems. All three systems can produce a very high vacuum and are all backed by a mechanical pump. The mechanical pump serves as a general workhorse for most mass spectrometers and allows for an initial vacuum of about 10^{-3} torr to be obtained. Once a 10^{-3} torr vacuum is achieved, the other pumping systems, such as diffusion, cryogenic and turbomolecular can be activated to obtain pressures as low as 10^{-11} torr. Maintaining a good vacuum on a mass spectrometer is critical to its overall performance.

Overview

The mass analyzer is a critical component to the performance of any mass spectrometer. Among the most commonly used are the quadrupole, triple quadrupole, quadrupole ion trap, time-of-flight, time-of-flight reflectron, quadrupole time-of-flight reflectron, orbitrap, and FTMS. However, the list is growing as more specialized analyzers allow for more difficult questions to be addressed. For example, the development of the quad-TOF has demonstrated its superior capabilities in high accuracy tandem mass spectrometry experiments. Once the ions are separated by the mass analyzer they reach the ion detector, which is ultimately responsible for the signal we observe in the mass spectrum.

3. Mass Analyzers

Questions

- What does a mass analyzer measure?

    Mass-to-charge ratio or m/z.

- What is tandem mass spectrometry?

    Tandem mass spectrometry is the act of exposing molecular ions to fragmentation via an inert gas thus enabling the creation of fragment ions that can be used to identify and/or quantify molecules.

- What is a general definition of resolution?

    $M/\Delta M$

- How does delayed extraction cool MALDI ions prior to mass analysis?

    Delayed extraction cools ions via collisions with inert gas molecules like nitrogen, thus reducing their velocities prior to ejection from the ionization source. This positively impacts resolution.

- What is the primary advantage of FTMS?

    FTMS instruments (including ion cyclotron resonance and orbitraps) generally have higher resolution than other mass analyzers.

- Why do mass analyzers need a vacuum?

    The vacuum minimizes ion collisions with neutral molecules. These collisions impact resolution and generate fragmentation.

- If you could only buy one mass spectrometer, which would it be?

    The most popular MS system is the ESI triple quadrupole.

- Who won a Nobel prize for developing quadrupole mass analyzers?

    Wolfgang Paul

Useful References

Busch K.L., Glish G.L., McLuckey S.A. Mass Spectrometry/Mass Spectrometry: Techniques and Applications of Tandem. John Wiley & Sons, **1989**.

Cotter R. Time-Of-Flight Mass Spectrometry: Instrumentation and Applications in Biological Research. Washington, D.C.: ACS, **1997**.

McCloskey J.A. & Simon M.I. *Methods in Enzymology: Mass Spectrometry*. Academic Press, **1997**.

Kinter M. & Sherman NE. Protein Sequencing and Identification Using Tandem Mass Spectrometry. Wiley-Interscience, **2000**.

Markarov A. Electrostatic Axially Harmonic Orbital Trapping: A High-Performance Technique of Mass Analysis. *Analytical Chemistry*, **2000**.

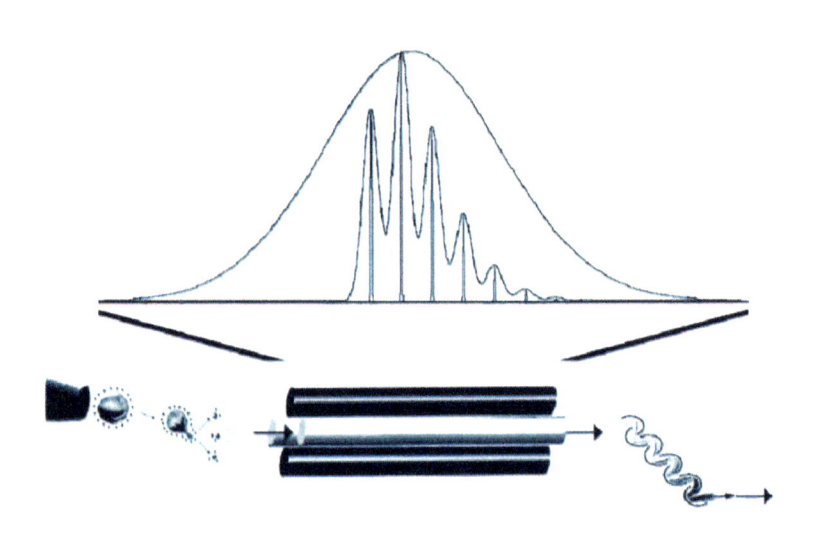

Chapter 4

Practical Aspects of Metabolite, Lipid and Peptide Analysis

Then and Now

This chapter examines some of the practical aspects (e.g., isotopes and resolution **Figure 4.1**) of using ESI and MALDI mass spectrometry for metabolites, lipids, peptides, and proteins. What is important for mass spectrometry today is in stark contrast to what was important in the past. In the 1960s and 1970s biomolecular mass analysis was typically accomplished by covalently adding protecting groups to make molecules more stable to thermal vaporization. ESI and MALDI largely changed this concept and ushered in a more comprehensive and sensitive era for biomolecule analysis.

protonated molecule (MH+)
$C_{81}H_{115}N_{24}O_{36}$

monoisotopic mass = 1999.7905

average mass = 2000.9200

nominal mass = 1999

resolution = 200
resolution = 2000
resolution = 20,000

m/z

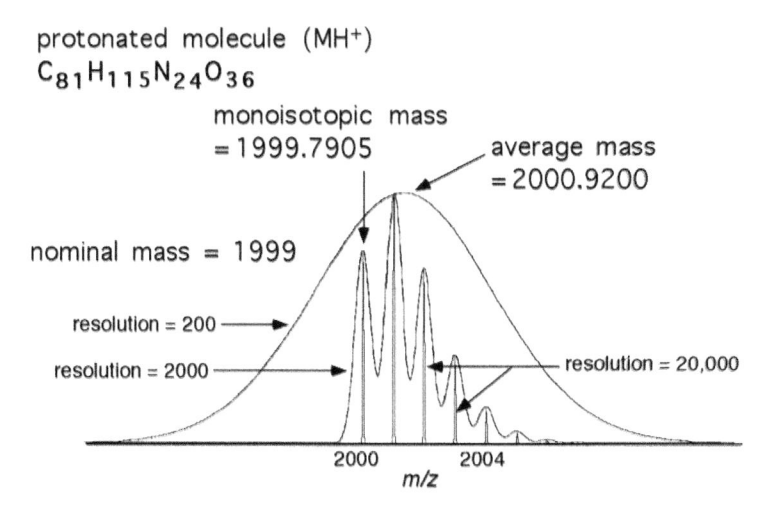

Figure 4.1. The mass spectrum of a protonated molecule obtained at resolving powers of 200, 2000, and 20,000 (using the FWHM definition of resolution).

This is an example of how the resolvingpower can have a dramatic effect on resolving isotopes.

During a transition period in the early 1980s the ionization technique called fast atom/ion bombardment was developed and successfully used to analyze peptides and other biomolecules. However, FAB was found to have relatively low sensitivity and generated ions within a limited mass range. In the late 1980s ESI and MALDI dramatically improved the sensitivity of mass spectrometry and the ability to generate ions of a wide variety of biomolecules. Yet with ESI and MALDI came new considerations for preparing samples to maximize sensitivity, resolution, and accuracy.

Some important characteristics and considerations of performing mass measurements today include quantitation, molecular weight calculation, isotope patterns, ionization characteristics, salt content, sample purity, calibration/accuracy, resolution, sensitivity, solubility, speed, matrix selection and matrix preparation.

Quantitation

One common question about ESI and MALDI is whether the ion intensities correlate to the relative amounts of each component. In most cases MALDI does not provide this type of information unless the compound has been calibrated against an internal standard. However, for ESI it is possible to get some quantitative information based on external calibration, although internal calibration still does provide the best accuracy. **Table 4.1** lists important factors that affect the ability to perform quantitative measurements. For instance, for compounds that have similar mass and functional groups, the relative ion intensities may correspond to their content. The ability of a molecule to become ionized is closely related to its functional groups. The role of functionality in ionization is demonstrated for two compounds, an amine, and an amide in **Figure 4.2**. Both are at the same concentration, yet they have significantly different signals because of the relatively high proton affinity of the amine. Another example is shown in **Figure 4.3** where excellent linearity is obtained between very similar compounds, the cyclic peptides cyclosporine A and G, because they differ only by a methylene group. Typically, the best quantitation is obtained when a compound is calibrated against an internal standard structurally similar to the molecule in question.

Table 4.1. Factors affecting quantitation.

Averaging	The signal to noise ratio (S/N) is directly proportional to the square root of the number of scans averaged. More averaging results in less errors associated with random noise. $S/N \propto N^{1/2}$
Amount of material	Signal intensity will fluctuate significantly at low sample amounts, reducing quantitative capability. MALDI typically provides a less stable signal than ESI and therefore generates bigger errors with less sample. Picomole quantities of material will generally provide good quantitation and with some instrumentation it is possible to go down to the low femtomole level.
Dynamic signal range	Analyzers, such as quadrupole ion traps, have a relatively small dynamic range ($\sim 10^3$-10^4) and can easily get saturated. Quadrupole analyzers have a larger dynamic signal range ($\sim 10^6$) and are therefore more well suited to quantitative measurements.
Ionization technique (instrument stability)	ESI has a relatively stable signal and as a result excellent quantitation can be achieved with a minimal amount of averaging. MALDI is less stable (from laser shot to laser shot) and requires greater care in obtaining quantitative data.
Compound's functional groups	The functional groups on a molecule can drastically affect the ionization properties. For instance, an amine will pick up a proton far more efficiently than an amide. Therefore, to obtain good quantitative data an internal standard with comparable ionization characteristics is desirable.
Choice of internal standard	An internal standard with comparable ionization characteristics to the compound of interest allows for consistent relative signal stability. The best choice is an isotopically labeled internal standard.

Consistent sample handling | Variations in sampling handling approaches (solvents, mixing, injecting...) can have a significant effect on the quality of quantitative results.

Figure 4.2. The ESI mass spectrum was obtained from an equal amount of each compound, illustrating that the ion intensity does not necessarily correlate to the amount of sample being analyzed.

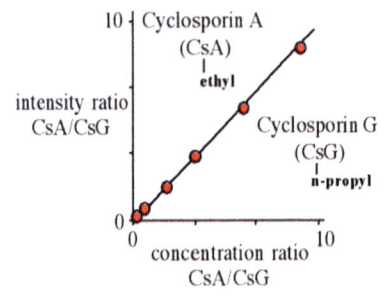

Figure 4.3. The quantitative analysis of cyclosporin A using cyclosporin G as an internal standard. These two molecules differ by a single methylene group which makes their ionization properties, for practical purposes, very similar. Cyclosporin G (0.20 µM) was used as an internal standard. Similar results were obtained with ESI and MALDI-MS.

A common calibration method used for endogenous metabolite and pharmacokinetic studies is "internal standardization" where a precise quantity of reference material is "spiked" into a sample. A requirement needed for internal standards is that their physicochemical characteristics should be identical or similar to those of the analyte of interest during the measurement. Traditionally, stable isotope labeled compounds and structural homologs, or analogs have been used as internal standards. When an internal standard is used, quantitation is typically based on a ratio of the analyte to the internal standard, multiplied by the known concentration of the internal standard (**Figure 4.3**). Generally, the more closely the final concentration of the reference material is to the analyte, the more reliable are the results.

This simple method of quantitation is acceptable if the linearity of the method, that is the response factor or intensity-of-signal per unit concentration, has been demonstrated to be constant over the concentration range of both the analyte and internal standard. While it is often challenging to arrive at an appropriate internal standard, the benefit of this standardization technique is that the internal standard can be added to a sample early in the analysis and prior to sample preparation. Therefore, if the internal standard and analyte have similar characteristics, any loss of the analyte during sample preparation will be reflected by a concomitant loss of internal standard but the ratio of their concentrations will still reflect the original quantity of the analyte. The benefits also extend to the analytical system, where the absence of internal standard peaks during a run can lead to accelerated identification of injector errors or other system failures.

Calibration of a method using an external standard has the benefit of allowing use of authentic reference material where the reference material is used to generate an external calibration curve for the batch of samples. The frequency of recalibration of the system depends on the stability of the analytical methods; on an LC-MS/MS quadrupole system, it can be as frequent as twice daily. In general, "system-suitability standards" or "QC/QA controls" are also interspersed in the sample batch during analysis. These controls serve as internal process controls that qualify the analytical testing throughout the entire process.

Calculating Molecular Weight

There are three different ways to calculate mass from the molecular formula: the average mass, the monoisotopic mass, and the nominal mass. Each calculation is used for specific reasons; for instance average mass calculation is used when the individual isotopes are not distinguishable (**Figure 4.1**, resolution = 200). The monoisotopic mass is calculated when it is possible to distinguish the isotopes (**Figure 4.1**, resolution = 2000). The nominal mass is not often used yet is typically applied to compounds containing the elements C, H, N, O and S and having a mass below 600 Da. Examples are shown in **Table 4.2**.

Monoisotopic mass (exact mass) -- The mass of an ion for a given empirical formula calculated using the *exact* mass of the most abundant isotope of each element (e.g., carbon-12).

Average mass -- The mass of an ion for a given empirical formula, calculated using the average atomic weight averaged over all the isotopes for each element.

Nominal mass -- The mass of an ion with a given empirical formula calculated using the *integer* mass of the most abundant isotope of each element.

A compound with a formula $C_{60}H_{122}N_{20}O_{16}S_2$ has a monoisotopic mass of 1442.8788, an average mass of 1443.8857 and a nominal mass of 1442.

Table 4.2. An illustration of the differences between monoisotopic, average, and nominal mass for some elements, a lipid, a sugar, and a peptide.

Name	Molecular Formula	Average Mass (Da)	Monoisotopic Mass (Da)	Nominal Mass (Da)
Hydrogen	H	1.0080	1.0078	1
Carbon	C	12.0112	12.0000	12
Nitrogen	N	14.0067	14.0031	14
Oxygen	O	15.9994	15.9949	16
Sulfur	S	32.0600	31.9721	32
A lipid	$C_{18}H_{35}N_1O_1$	281.4858	281.2718	281
A sugar	$C_{56}H_{118}N_4O_{14}$	1071.5833	1070.8644	1070
A peptide	$C_{101}H_{258}N_{24}O_{24}$	2193.3288	2191.9704	2190

One of the most important features of a mass spectrometer is resolving power (Chapter 2) which turns out to also be important when calculating the mass of a molecule. For example, if the isotopes can be distinguished in the mass spectrum, the observed *m/z* will correspond to the monoisotopic mass. This is illustrated in

Figure 4.1 where at a resolution of 2000 on an electrospray quadrupole instrument there is sufficient resolving power to distinguish isotopes on a peptide. FTMS instruments, because of their high resolving power, can provide isotopic distributions of proteins with mass above 10,000 Da. Generally speaking, the average mass is a satisfactory calculation for compounds above a mass of 5000 Da while for compounds below 5000 Da, the resolution of the instrument is an important factor in determining which mass calculation to use.

Isotope Patterns

Figure 4.1 illustrates how the isotope pattern of an individual molecule, as well as the resolving power of the mass spectrometer, can impact the mass spectrum. A mass spectrometer with a resolving power of 2000 can resolve the isotopes for ions of at least 2000 Da. **Figure 4.1** also demonstrates the effect isotopes can have on the observed molecular weight. Compounds containing ^{12}C, ^{1}H, ^{14}N, and ^{16}O have isotopes ^{13}C, ^{2}H, ^{15}N, ^{17}O, and ^{18}O in relatively low abundance (1.10%, 0.015%, 0.366%, 0.038%, and 0.200%, respectively) yet these less abundant isotopes make a significant contribution when enough atoms are present.

The isotope patterns are often taken for granted, but they can be a great source of information. For instance, the spacing of the isotopes can tell you about the charge state (e.g. 1/2 spacing = 2^+ charge state, 1/3 spacing = 3^+ charge state). In addition, certain elements have distinct isotope patterns such as chlorine and bromine. Looking at the isotopic patterns in **Figure 4.4**, note that a compound containing one chlorine atom will have an isotopic contribution that correlates to a peak having a mass at **M + 2**, with a height of ~32% of the primary ion. The percentage contribution for chlorine and bromine will change if more of these halogens are added to the compound. Observing these isotopes can be a useful confirmation of the presence of these elements.

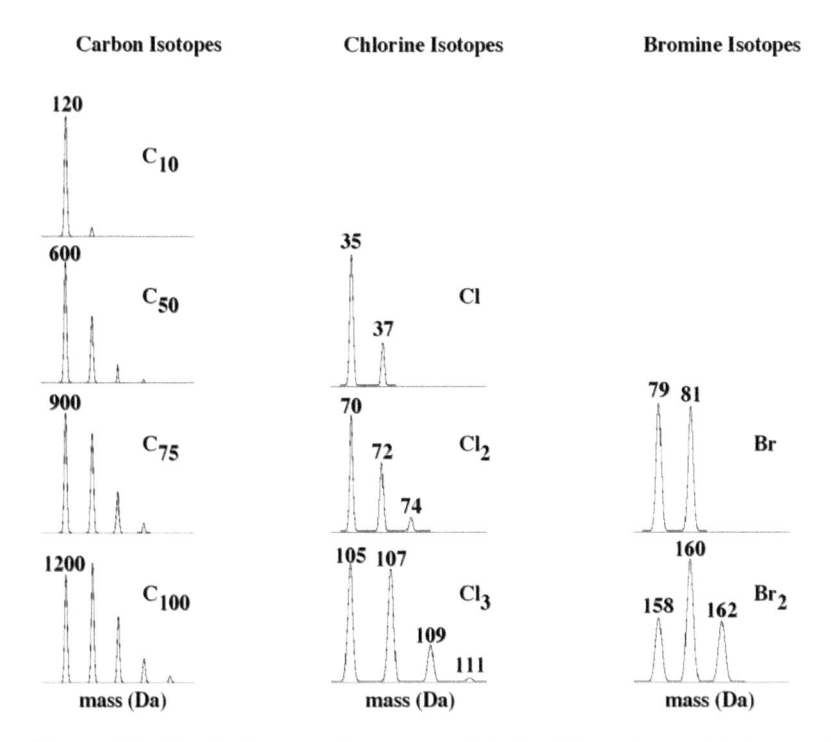

Figure 4.4. The isotope patterns associated with carbon, chlorine and bromine. The chlorine and bromine isotopes produce characteristic patterns that help one readily identify these elements in a compound. For carbon the natural contribution of 1.1% of the ^{13}C accounts for the isotopic abundance of ^{13}C at higher numbers of carbons.

Below 3000 Da the most common isotopic distribution that you will observe for biological compounds is dominated by carbon (**Figure 4.4**) since less abundant isotopes of hydrogen, oxygen, and nitrogen do not significantly contribute to the distribution below this mass. However, sulfur atoms can present a significant contribution as ^{34}S is abundant at 4.68%.

Solubility

Sample solubility is critical to obtaining quality data. For instance, if you are performing ESI on a sample that has precipitated out of solvent, you will not observe that compound. The solvent or matrix is the medium that allows your sample to be transported to the gas

4. Practical Aspects of Metabolite, Lipid and Peptide Analysis

phase. Moreover, solvents often plays an important role in ionization. For techniques like ESI and APCI, if your compound is not soluble in the solvent, it will be impossible to observe a signal. **Figure 4.5** illustrates the dramatic effect that choosing a suitable solvent can have on acquiring data.

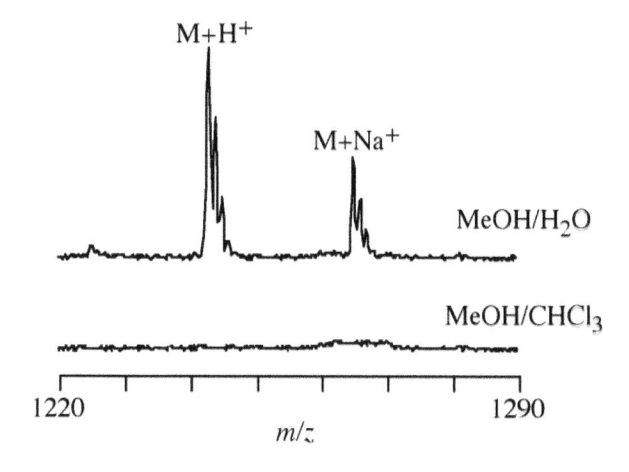

Figure 4.5. Electrospray mass spectra of a peptide analyzed from methanol/water (top) and in methanol/chloroform (bottom). In this example the peptide was not observed in the methanol/chloroform solvent system.

Timing

Analyzing the sample as soon as possible after it has been prepared is important, as it is quite common for compounds to decompose, react with the solvent in a relatively short time, or enzymatically degrade. A problem associated with hydrophobic compounds is loss of the sample to the container's surface. Another phenomenon that has been observed is that a sample solution may even leach some salts or plasticizers from the container. Compounds may also stick to the side of vials, therefore the choice of containers is also important.

Calibration and Mass Accuracy

Accuracy is one of the most important aspects of the data obtained from a mass spectrometer. It is important for compound identification and more recently, for high accuracy measurement of proteolytic peptide fragments for protein identification. For high accuracy to be possible one needs an instrument with resolution on the order of ~4000 or higher. For very high accuracy (<5 ppm) it is often necessary to have an internal standard present or at least some reference compound.

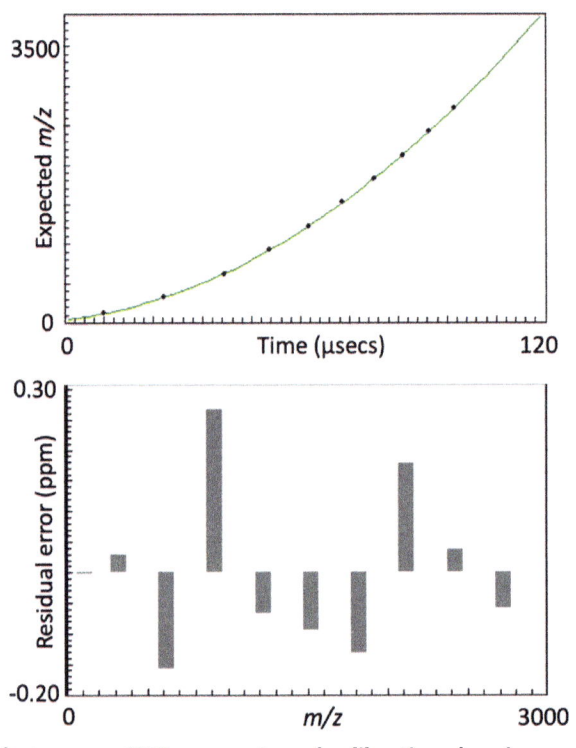

Figure 4.6. An electrospray TOF mass external calibration showing excellent accuracy (typically <0.5 ppm) of the calibration ions.

Standard compounds are used to calibrate a mass spectrometer's mass analyzer with respect to how it measures the m/z (**Figure 4.6**). Standards are also used internally (**Figure 4.7**) while running an unknown sample to improve accuracy. Calibration is generally performed using a standard mixture that generates a reliable source of known ions that cover the mass range of interest. For instance, polypropylene glycols (PPG's) have been used as electrospray calibrants. To remove the ambiguity associated with the isotope peaks generated with polypropylene glycols, cesium iodide (CsI) has also been used for calibration or fluorinated phosphonium salts (**Figure 4.6**). Fluorinated compounds are useful because monoisotopic fluorine minimizes the isotope contribution and therefore makes the signals simple to interpret. When a calibration compound of known mass is measured, it helps determine whether the instrument is running within acceptable error limits. If the instrument is off in accuracy, an adjustment based on the calibration compound will allow the instrument to run accurately again.

It is important to distinguish between internal and external calibration. External calibration refers to the instrument being calibrated followed by analysis without the presence of a calibrant. It is important to keep instrument and solutions free from contamination. The instrument can mistake impurities that are close in mass to the calibration compounds and cause the instrument to calibrate incorrectly.

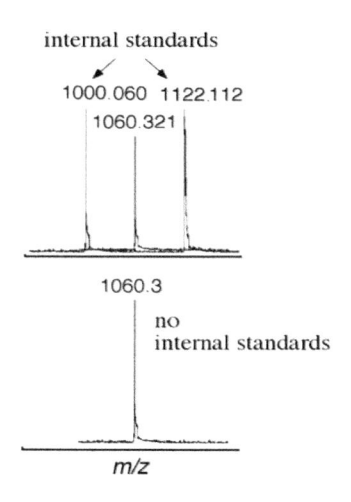

Figure 4.7. The internal standards can be used to obtain high accuracy when used with high resolution instruments. In this case the spectral data containing the internal standards allows us to get accuracy to the 5-ppm level or better.

Internal calibration refers to analyses that are performed with a calibrant present and serves to further improve accuracy. "Acceptable" levels of accuracy are highly dependent on the mass analyzer; for a quadrupole ~200 ppm is acceptable; for FTMS ~5 ppm is acceptable with external calibration. Internal calibration,

~100 ppm for quadrupoles and <5 ppm for FTMS and TOF analyzers is acceptable. The improvements in accuracy for TOF analyzers include making the flight tube from material that has a low thermal expansion coefficient (the flight tube does not change appreciably as a function of temperature). The calibration of MALDI TOF and ESI quadrupole mass spectrometers often use the standard mixtures shown in **Table 4.3**.

Table 4.3. Common MALDI and ESI calibration compounds and *m/z* values.

Calibrants	MH⁺	MH₂²⁺	MH₃³⁺
Matrix-Assisted Laser Desorption/Ionization (MALDI) Calibrants			
Angiotensin II	1046.54 (mono*)		
	1047.20 (avg)**		
ACTH (18-39)	2465.20 (mono)		
	2466.73 (avg)		
Insulin bovine	5734.56 (avg)	2867.78 (avg)	
apo-Myoglobin equine	16952.47 (avg)	8476.74 (avg)	
Cytochrome C equine	12,361.09 (avg)	6181.05 (avg)	
BSA	66,430 (avg)	33216 (avg)	22144. (avg)
BSA-dimer	132,859 (avg)	66430 (avg)	44287. (avg)
Electrospray Ionization (ESI) Calibrants			
Sodium Formate	100-1100 *m/z*		
PPG	100-3000 *m/z*		
CsI	100-3000 *m/z*		
Agilent calibrant ($C_xH_yO_zNa_pP_bF_c$)	100-3000 *m/z*		

*mono is monoisotopic mass   **avg is average mass

Sample Purity and Clean Containers

Whether analysis is being performed with MALDI or ESI, sample (and solvent) purity maximizes sensitivity. It is true that salts can facilitate ionization for certain compounds such as carbohydrates, however, in general, excessive salt and other contaminants will lead to reduced sensitivity. An effective approach for higher sensitivity ESI-MS is reverse phase liquid chromatography since it separates compounds prior to analysis. Two common methods (**Figure 4.8**) of cleaning samples for MALDI-MS analysis involve **ZipTip**™ or cold water washing, both of which are effective

4. Practical Aspects of Metabolite, Lipid and Peptide Analysis

for salt removal. Another way we reduce salts on the spot and enhance signal is doing a 1x dilution of the sample on the plate by spotting 0.25uL sample, 0.25uL water and 0.5uL matrix. This seems to especially reduce buffer effects for proteins submitted in buffers such as PBS that can cause suppression.

Figure 4.8.

ZipTip™ pipette tip containing C18 material

- ZipTip™ is placed on standard pipetter;
  - sample is aspirated onto the C18 material;
  - is washed and

eluted. On-plate

MALDI wash

- add 1µl of sample to plate and allow to dry;
- add matrix;
- allow to dry;
- add 2µl of H_2O, allow to sit for 10 seconds, remove.

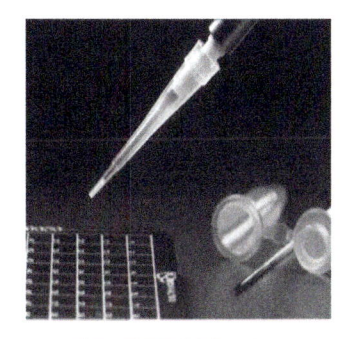

Generally, the more pure or homogenous a compound is, the better the sensitivity and quality of the mass spectral data. For example, electrospray requires that the sample be relatively free (less than 10 millimolar concentration) of salts, buffers, and other contaminants because these impurities have a detrimental effect on the electrosprayed droplets and their ability to evaporate. If a droplet cannot effectively evaporate because of reduced vapor pressure associated with impurities; ion production will be lost. Another effect of contamination is the loss of sensitivity due to the contaminants' competition for protons/cations in the droplets. Using liquid chromatography as a sample introduction method for ESI-MS is very effective for generating pure samples and therefore maximizing signal.

Even though MALDI is known to be more tolerant of salts, buffers, and impurities, cleanup procedures are still useful. With MALDI, one typically probes the surface with the laser beam until a reasonable signal has been found. A good signal is generated because of the laser beam striking a portion of the probe that has the sample embedded in crystalline matrix. To facilitate sample incorporation and desalting, the sample can be washed clean using a droplet of cold water (**Figure 4.8**). This method promotes a greater and more uniform ion signal.

Table 4.4 presents an overview of the effect different salts and detergents can have on ESI and MALDI. Tris-HCl (25-50mM) and ammonium bicarbonate (25-50 mM) are considered the most compatible buffers with mass spectrometry. Conveniently, these buffers are also compatible with enzymes such as Arg-C, Trp, Lys-C, Chymo, and Glu-C.

It can go without saying that containers such as glass or the eppendorf should be clean, however sample containers cleanliness can vary widely from manufacturer to manufacturer. Therefore, it is another source of potential contamination to consider when problem solving. For example, plastic vials leech contaminants if using organic solvents and can therefore interfere with experiments.

4. Practical Aspects of Metabolite, Lipid and Peptide Analysis

Table 4.4. Maximum surfactant & buffer concentrations for MALDI & ESI.

Surfactant, Buffer & Salts	MW (g/mol)	MALDI (mM)	MALDI (wt.%)	ESI (mM)	ESI (wt.%)	Reference
TRIS	121	100	1.0	n. a.	n. .	A, B
HEPES	238	100	2.4	n. a.	n. a.	A, B
BICINE	163	50	0.8	n. a.	n. a.	B
Urea	60	500	3.0	n. a.	n. a.	C, D
Guanidine, HCl	96	250	2.4	n. a.	n. a.	C, D
Dithiothreitol	154	500	7.7	n. a.	n. a.	D
Glycerol	92	130	1.2	n. a.	n. a.	C, D
N-Octyl-glucopyranoside	292	3.4	0.1	3.4	0.1	C, E
n-Octyl sucrose	468	n. a.	n. a.	2.1	0.1	E
n-Dodecyl sucrose	524	n. a.	n. a.	1.9	0.1	E
n-Dodecyl maltoside	511	n. a.	n. a.	2.0	0.1	E
Octyl thioglucoside	308	n. a.	n. a.	3.2	0.1	E
n- Hexyl glucoside	264	n. a.	n. a.	3.8	0.1	E
n-Dodecyl glucoside	348	n. a.	n. a.	2.9	0.1	E
PEG1000	1000	n. a.	n. a.	0.5	0.05	F
PEG2000	2000	0.5	0.1	n. a.	n. a.	C,
Triton X-100	628	1.6	0.1	<1.6	<0.1	C, E
NP-40	603	1.7	0.1	n.a.	n. a.	
Zwittergent, 3-16	392	2.6	0.1	n. a.	n. a.	C
Tween20	1228	n. a.	n. a.	0.81	0.1	E
Thesit	583	n. a.	n. a.	<1.7	<0.1	E
SDS	288	0.35	0.01	0.335	0.01	C, D, E, F
LDAO	229	4.4	1.0	<4.4	<0.1	C, F
CTAB	284	n. a.	n. a.	<3.5	<0.1	F
CHAPS	615	0.16	0.01	1.6	0.1	C, E
Sodium Cholate	431	n. a.	n. a.	2.3	0.1	E
Sodium Taurocholate	538	n. a.	n. a.	<1.9	<0.1	F
Sodium Azide	65	15	0.1	3.1	0.02	C, F
NH$_4$HCO$_3$	79	50	0.4	n. a.	n. a.	D
NaCl	58	50	0.29	n. a.	n. a.	C, D
Sodium Acetate	82	50	0.41	n. a.	n. a.	B, C
NaHPO$_4$	120	10	0.12	10	0.12	B, C, D, F
TFA	114	n. a.	n. a.	4.4	0.05	Pri. Comm.

HEPES: N-[2-hydroxyethyl]piperazine-N'-[2-ethanesulfonic acid]
TRIS: Tris[hydroxymethyl]aminomethaneacetate
BICINE: N,N-bis[2-hydroxyethyl]glycine
TFA: Trifluoroacetic acid
CHAPS: 3-[(3-cholamidopropyl)-dimethylammonio]-1-propanesulfonate
PEG: polyethylene glycol
LDAO: Lauryldimethylamine oxide

References:
A: Kallweit U. et al., Rapid Comm. Mass Spec. 10, 845-849, 1996.
B: Yao J. et al., J. Am. Soc. Mass Spectrom 9, 805-813, 1998.
C: Coligan J. E. et al., In Current Protocols in Protein Science 2, 16.2.
D: Gevaert K. et al., ABRF web publication, 1998.
E: Ogorzalek et al., Protein Science 3, 1975-1983.
F: Kay I. and Mallet A. I. Rapid Comm. Mass Spec. 7, 744-746, 1993.

Sensitivity / Saturation

Too little or even too much sample will have a dramatic effect on the data and can make the difference between a successful and an unsuccessful analysis. For example, if too little sample is used, understandably the instrument will be unable to detect a signal. If too much is used it can skew the intensity profile and mass accuracy of the ions observed (**Figure 4.9**). High concentrations can also make impurities appear to be more dominant or even cause signal suppression. An important point is that higher concentration is not always better and that it is more important to be within the correct range for your instrument.

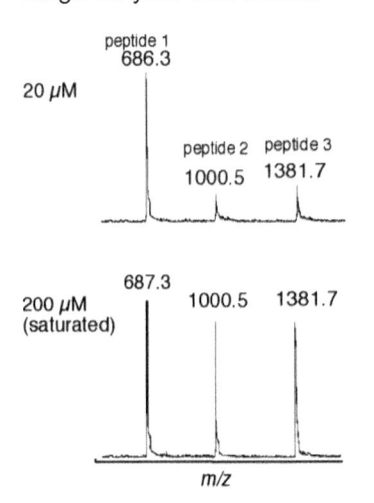

Figure 4.9. The electrospray mass spectra of three peptides is shown. At 20 µM concentration of peptide 1 the two other peptides can be observed. However, at a 10-fold higher concentration (200 µM) the relatively minor impurities, peptides 2 and 3, now appear as major components due to detector saturation of peptide 1. Peptide 1 is so saturated that the C13 isotope of this peptide appears to be bigger than the 12C.

But what is the "right" range? The answer is highly dependent on the instrument, sample, and technique. Some instruments of very similar design can vary widely in terms of their dynamic range. However, **Table 4.5** provides some general guidelines for the amount of sample required according to the ionization technique. **Table 4.5** converts a range of concentrations often used in electrospray and MALDI from micromolar (10-100 µM) to milligrams/milliliter with respect to molecular weight. Oligonucleotides for example generate better signal at the 100uM range versus peptides which typically ionize excellently even at the 10uM range, or lower. Those used for FAB analysis are also displayed. The concentrations shown are conservative; most instruments will easily handle significantly lower concentrations. In

4. Ionization Characteriscs

general, however, the concentrations listed will produce good signal even in the presence of some salt or other contaminating compounds. In comparing ESI, MALDI, and FAB, **Table 4.5** illustrates the improvements that ESI and MALDI offer and why they are so important today.

Table 4.5. Typical concentration range for ESI and MALDI (with conversion to µg/µl) as it compares to traditional FAB technology.

Molecular Weight	MALDI or ESI (1 nanomolar to 50 micromolar)	FAB (1 millimolar)
500 Da	0.0000005 - 0.025 µg/µL	≥ 0.50 µg/µL
1000 Da	0.000001 - 0.05 µg/µL	≥ 1.00 µg/µL
5000 Da	0.000005 - 0.25 µg/µL	≥ 5.00 µg/µL
10,000 Da	0.00001 - 0.50 µg/µL	≥10.00 µg/µL
50,000 Da	0.00005 - 2.50 µg/µL	not applicable
100,000 Da	0.0001 - 5.00 µg/µL	not applicable

* Typical sample volumes used for MALDI, ESI & FAB are 1, 10, and 2µL, respectively.

Ionization Characteristics

The types of functional groups on a molecule will often determine how a compound should be analyzed. When analyzing a compound by ESI or MALDI, first look for sites on the molecule that can be ionized. Amines (+), acids (-), and amides (+) represent easily ionizable functional groups, while hydroxyl groups, esters, ketones, and aldehydes do not accept a charge as easily and are therefore more difficult to ionize, typically resulting in weak ion signals. Due to the presence of amide and amine groups, peptides usually ionize easily through protonation. Because peptides often contain amino acid residues with side chain carboxyl groups, they can also be observed in the negative ion mode by deprotonation. Some carbohydrates will accept a proton because of the presence of an amide bond, but in general they form stable ions upon the addition of a cation other than a proton, such as Na^+ or K^+ to form $M + Na^+$ and $M + K^+$ ions. Oligonucleotides, proteins, and the myriad of small molecules each have their ionization peculiarities, making the conditions under which analysis is performed all the more important.

Table 4.6 describes some of how different compounds are analyzed. **Table 4.6. Typical ionization properties for biomolecules.**

Compound	Ionization Mechanism	MALDI & ESI
Peptides	Protonation/Deprotonation	Positive/Negative
Proteins	Protonation	Positive
Membrane Proteins	Protonation	Positive
Glycoproteins	Protonation	Positive
Carbohydrates & Protected Carbohydrates	Protonation/Cationization/Deprotonation/ Anionization	Positive/Negative
Oligonucleotides	Protonation/Deprotonation	Positive/Negative
Protected Oligonucleotides	Deprotonation/Protonation/Cationization	Positive/Negative
Small Biomolecules	Protonation/Cationization/Deprotonation Anionization ElectronEjection Electron Capture	Positive/Negative

MALDI Matrix Selection and Preparation

The choice of matrix is critical for MALDI and, in many cases, will determine the quality of the data (**Figure 4.10**). The introduction of new matricies has complicated matrix selection and preparation, yet at the same time had a tremendous impact on sensitivity and accuracy.

Sample preparation with MALDI involves dissolving the sample into a solvent, such as water, acetone, or methanol. The sample solution is then added to the matrix, commonly a nicotinic acid or cinnamic acid derivative, at a ratio of approximately 1 part sample to 10,000 parts matrix molecules. Approximately 10 picomoles (1 µl of a 10 µM solution) of the sample is added to an equal volume of a saturated solution of the matrix. It is important to note that MALDI most commonly involves a solid matrix and, therefore, MALDI sample preparation requires that after the sample solution is made up and deposited upon the probe, it is necessary to wait a few minutes for complete evaporation of the solvent to form nice crystals and leave a solid mixture of the sample embedded in the matrix.

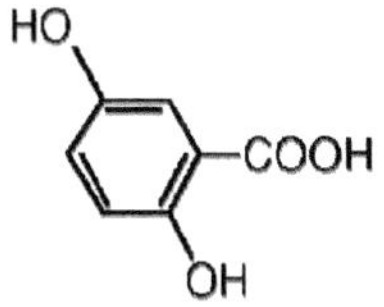

4. Practical Aspects of Metabolite, Lipid and Peptide Analysis

2,5-dihydroxy benzoic acid (DHB)
peptides, carbohydrates, synthetic polymers, and glycolipids

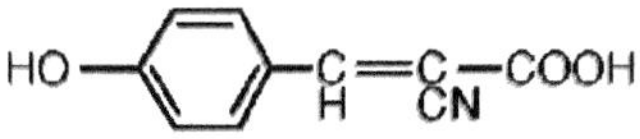

α-cyano-4-hydroxycinnamic acid (CCA)
peptides and proteins

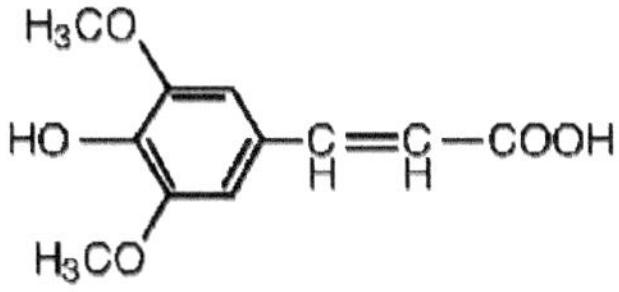

3,5-dimethoxy-4-hydroxycinnamic acid (sinapinic acid) peptides and proteins

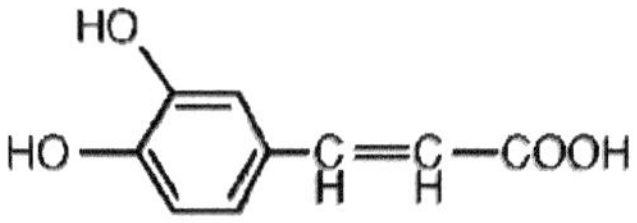

3,4-dihydroxycinnamic acid (ferulic acid)
peptides, proteins, and some oligos

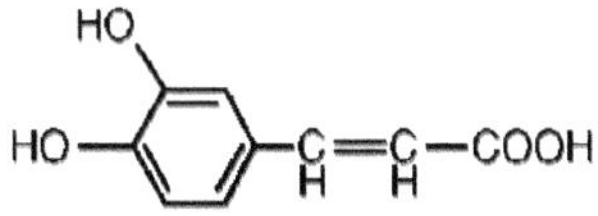

3,4-dihydroxycinnamic acid (caffeic acid)
peptides, proteins, and some oligos

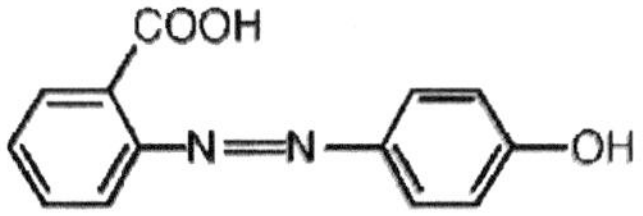

2-(4-hydroxyphenylazo)-benzoic acid (caffeic acid) (HABA)
large proteins, polar and nonpolar synthetic polymers

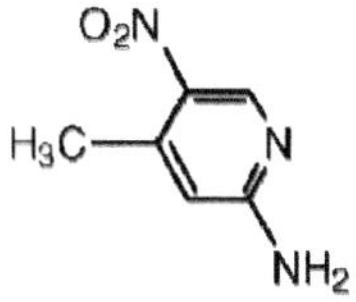

2-amino-4-methyl-5-nitropyridine
small acid-sensitive proteins (<12,000 Da)

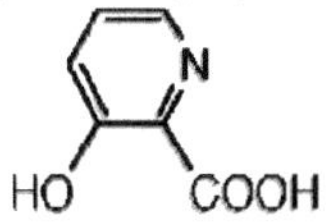

3-hydroxypicolinic acid (HPA) with ammonium citrate additive
oligonucleotides

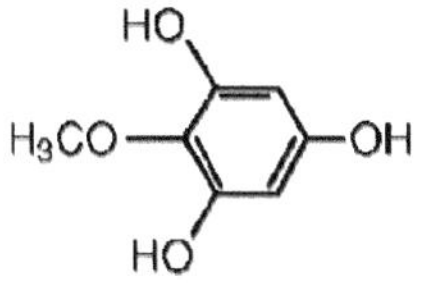

2,4,6-trihydroxy acetophonone (THAP) with ammonium citrate additive
oligonucleotides

Figure 4.10. Cinnamic acid is the basis of many of the common MALDI matrices. Sample deposition involves placing the sample and matrix onto a probe at a ratio of roughly 10,000 matrix molecules to every sample molecule. The type of matrix is important with regard to the type of compound being analyzed.

The evaporation process for MALDI can be expedited by flowing a stream of nitrogen or air over the dissolved matrix or by using a volatile solvent (acetone) if possible. Evaporation can also be expedited by spotting smaller volumes or using concentrating plates like anchor chips. Multisampling probes also quicken the analysis time by allowing for many samples to be prepared at once.

Overview

This chapter examines many of the factors that allow for quality mass spectral information from ESI and MALDI mass spectrometers. Among the most important of these factors for obtaining signal of high quality are sample solubility, matrix selection, the ionization characteristics, salt content and purity. Factors that contribute to data interpretation include the isotope patterns, saturation effects and accuracy.

Common Questions

Question	Answer
1) What concentration of material do you need to perform a routine mass measurement?	~1-50 µM for electrospray ~1-50 µM for MALDI ~1000 µM for FAB
2) If you have the appropriate concentration, what volume of solution do you need?	~10 µl for electrospray ~1 µl for nanoelectrospray ~1 µl for MALDI ~1 µl for FAB
3) In a typical mass spectrometry experiment what would be the maximum mass range expected from each instrument?	~200,000 Da by electrospray ~300,000 Da by MALDI ~7000 Da by FAB

4. Practical Aspects of Metabolite, Lipid and Peptide Analysis

Question	Answer
4) What does resolving power mean in mass spectrometry?	Resolving power is the ability of a mass spectrometer to separate ions of similar mass-to-charge ratio (m/z). High resolution enables the analyst to resolve isotopic peaks of high-mass ions. Most commonly, resolving power is defined as M/ΔM as shown below and **Figure 4.3**. protonated molecule (MH+) $C_{81}H_{115}N_{24}O_{36}$ monoisotopic mass = 1999.7905 average mass = 2000.9200 nominal mass = 1999 resolution = 200 resolution = 2000 resolution = 20,000 2000 2004 m/z Resolving power = M/ΔM = 500/.5 = 1000
5) Are there always matrix peaks in the MALDI mass spectra?	Most often below m/z 500. It is possible for certain analytes to adjust the analyte-to-matrix ratio to minimize the matrix interference effect.
6) Do you observe matrix peaks with electrospray?	No. Since electrospray does not use a matrix, it does not generate matrix peaks. However, it does use solvents like water, methanol, and acetonitrile and it is possible to observe aggregates of these. At high salt concentrations it is also possible to see salt clusters and the observation of nonspecific noncovalent interactions.
7) Why are cation adducts predominantly observed in the mass spectrum of some molecules?	Some compounds do not form stable protonated molecules because the charge from the proton can be transferred onto the molecule and can thus destabilize it. Alkali cations can complex to some molecules because the charge remains localized and does not transfer onto the molecule (there is no charge destabilization). Therefore, the alkali cation adducts of some molecules are more readily observed than the protonated species.
8) How can you tell if a peak in electrospray is singly- or multiply- charged?	The spacing in isotope pattern will tell you what the charge state is. For instance, if the isotopes are spaced one m/z apart, the peak corresponds to a singly charged ion. If the isotopes are one half m/z apart then the peak corresponds to a doubly charged ion. If the peaks isotopes are spaced one third of a m/z apart then the peak corresponds to a triply charged ion. And so on.

Common Questions

Question	Answer
9) Electron ionization (EI) used to be the most popular ionization source, why is it not used as much anymore?	EI requires thermal desorption to vaporize a compound and many compounds will not survive the high temperature required for vaporization. Also, the electron ionization process is a relatively harsh ionization, and many molecules will not remain intact once ionized in this way.
10) Why does having too much salt in your electrospray sample cause signal suppression?	Electrospray ionization operates by the evaporation of charged droplets. Too much salt causes vapor pressure to go down and evaporation is thus inhibited.
11) Why do heterogeneous mixtures reduce the sensitivity of MALDI and electrospray?	MALDI and electrospray are both somewhat competitive ionization techniques in the sense that molecules in solution or the matrix will compete for protons. Therefore, the compound with the highest proton affinity will often dominate the spectrum. This tends to be more the case for electrospray since the molecules are competing in solution, while for MALDI many of the molecules are separated in the matrix solution, therefore reducing competition.
12) How should volatile samples be analyzed?	Volatile samples typically require more traditional ionization sources such as electron ionization.
13) How important is it to know the structure of a compound?	Structural information is useful in helping the analyst decide how to perform analyses. For example, an acidic molecule would be analyzed in negative ionization mode.
14) Why is it so important to minimize fragmentation during ionization?	The problem with obtaining fragment ions is that you can never be sure that the largest ion that is observed is really the molecular ion and not a fragment of something bigger. This is why ESI and MALDI are so useful; in many cases the ion generated corresponds to the intact molecular ion.

Chapter 5

Proteomics

Proteomics is very different from metabolomics, yet given their close technological association, I decided to cover this very broad topic in a chapter with the hope of covering some of the fundamental aspects. First, I will start with peptide and protein analysis, followed by protein identification via mass mapping, and then protein structural analysis. Although on some level they are all intertwined.

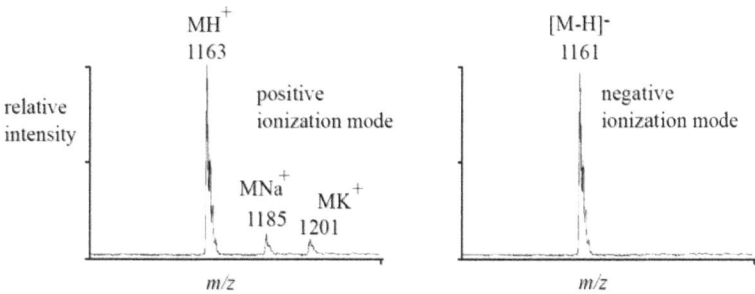

Figure 5.1. Electrospray ionization MS of a peptide in positive and negative ionization modes.

Peptide and Protein Analysis

Peptide and protein analysis with electrospray ionization (ESI) and matrix-assisted laser desorption/ionization (MALDI) is a very straightforward way to determine molecular weight whether you are interested in confirming an identity or in the initial stages of characterizing the structure. For example, what has made the mass analysis of peptides and proteins possible is the ability to promote their nondestructive vaporization/ionization with ESI and MALDI via the addition or removal of protons such as $[M + H]^+$, $[M + 2H]^{2+}$, $[M + nH]^{n+}$ or $[M - H]^-$, $[M - nH]^{n-}$ (**Figure 5.1**). During the generation of these species it is fortunate that typically little to no fragmentation is observed by either ESI or MALDI for peptides and virtually no fragmentation is observed for proteins. As a result, the generation of molecular weight information is often unambiguous.

In general, the sensitivity of mass spectrometry is quite good for proteins and peptides. It is possible to observe peptides at the attomole level or lower, while the sensitivity for protein analysis is typically not as good largely because of the protein's larger size. Proteins strike the detector with lower velocity than smaller molecules, resulting in less signal output. Even so, the sensitivity of MALDI and ESI for proteins is still considered good at the picomole to femtomole level.

In addition to molecular weight determination (**Figure 5.2**), MALDI and ESI have demonstrated increasing utility in the identification of proteins and protein post-translational modifications. While complete and routine sequence determination through mass analysis of proteins is yet to be realized, it is now possible to use proteolytic peptide fragments in combination with database searching algorithms to identify proteins. This is accomplished by combining mass spectrometry with enzymatic or chemical digestion (**Figure 5.2 insert**) followed by mass analysis of the peptide products and database searching techniques (especially useful in the post genome age).

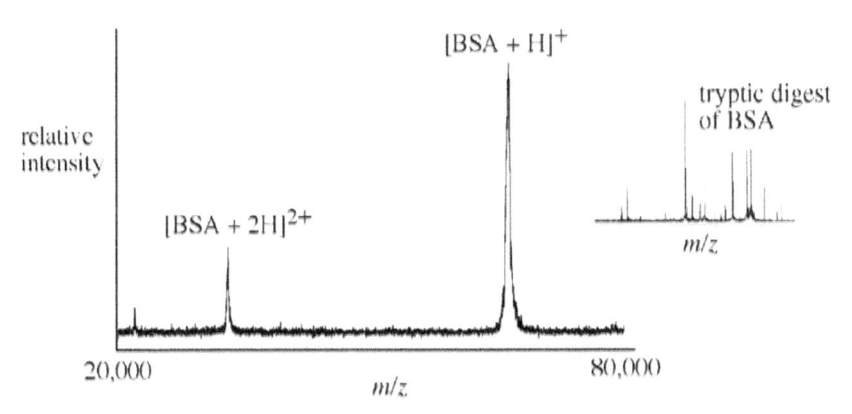

Figure 5.2. A typical MALDI mass spectrum of the protein bovine serum albumin (BSA) with inset of the tryptic digest of BSA.

MALDI and ESI mass spectrometry has also demonstrated its utility in identifying post-translational modifications, as these modifications usually lead to a predictable change in molecular weight. In addition, electrospray with tandem mass spectrometry is demonstrating a capacity for performing *de novo* sequencing.

Peptides and Proteins by MALDI

MALDI is useful for peptide and protein analysis for the following reasons:

- MALDI allows for high throughput analysis (**Figure 5.3**)
- high sensitivity (in the femtomole range or better)
- tolerance of heterogeneous samples (**Figures 5.4**)
- typically high accuracy mass measurements can now be generated on a MALDI-TOF reflectron in the mass range below 3000 *m/z* or higher with a MALDI Orbitraps and FTMS
- good dynamic range (**Figure 5.5**)

Factors in the automation of MALDI

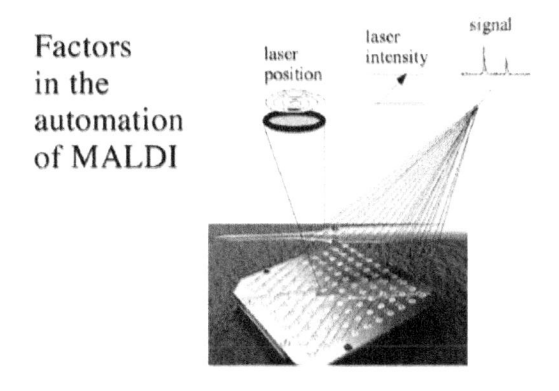

Figure 5.3. MALDI can employ multisample preparation and automated sample analysis. Three different parameters can be used to adjust and monitor autosampling, including laser position, laser intensity, and signal intensity.

Surface-based ionization techniques like MALDI allow for rapid analyses primarily because both allow for the preparation of multiple samples simultaneously using multi-sample probes. The automation of those analyses is becoming increasingly important in proteomics and combinatorial chemistry. These analyses are driven by a computer- controlled procedure to monitor the ion signal as a function of laser position and laser intensity (**Figure 5.3**). To accomplish this, the computer workstation can automatically adjust laser intensity and searches the sample well until a signal (within the specified mass range and intensity threshold) is obtained. Based on a careful preselection of autosampler options, each parameter (laser intensity, search pattern, step size in well, signal intensity, and m/z range) is adjusted to minimize time of analysis and maximize signal quality.

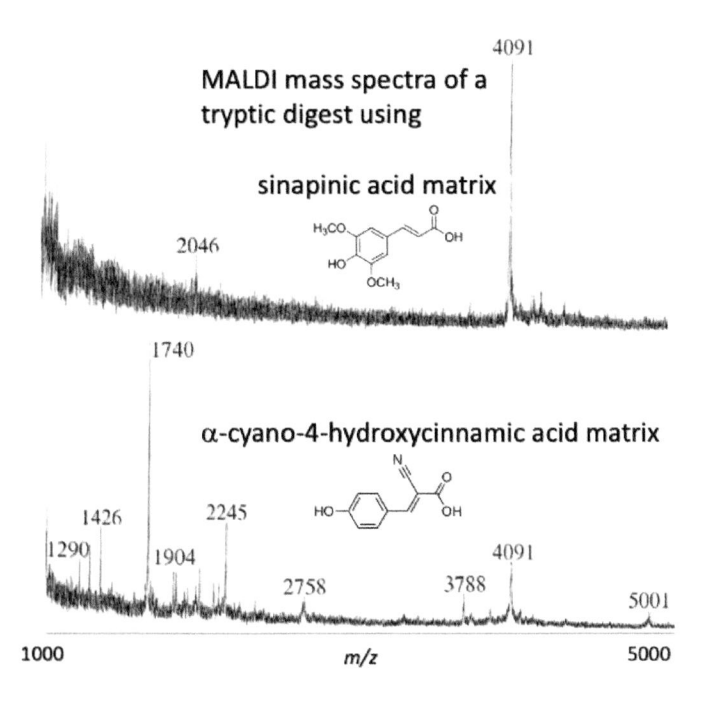

Figure 5.4. MALDI-MS analyses of the same tryptic digest of a protein using sinapinic acid and α-cyano matrices. Different matrices can provide significantly different MALDI results on depending on the sample type.

For MALDI analysis of relatively small molecules and peptides (400 Da - 1000 Da), 2,5-dihydroxybenzoic acid (DHB) matrix produces less interference than other matrices. Matrices such as α-cyano-4- hydroxycinnamic acid (α-cyano) give a high matrix background although they are particularly good for generating ions above 700 Daltons. For peptides generated by enzymatic digestion, α-cyano is often the matrix of choice. 3,5-dimethoxy-4-hydroxycinnamic acid (sinapinic acid), another common matrix, is most commonly used for whole protein analysis. Most matrices reported to date have been acidic, but basic matrices have also been introduced, such as 2-amino-4-methyl-5-nitropyridine and trihydroxyacetophenone (THAP), which extends the utility of MALDI to acid-sensitive peptides, proteins, or other acid-sensitive compounds. Interestingly, different matrices can provide dramatically different results on the same sample (**Figure 5.4**).

MALDI Sample Preparation Procedures Whole

proteins

- A saturated solution of sinapinic acid can be prepared in acetonitrile, water and TFA (50/50 acetonitrile/water with 0.1% TFA) or acetonitrile/water/TFA=1:1:0.1 and vortexed for complete mixing. The solution is ready for use when the undissolved solid settles.
- Place 0.5 μL protein solution (1-10 pmol/μL) directly on plate and then add 0.5 μL matrix solution.
- Perform on-plate wash with 2 μL nanopure H$_2$0 when sample is salty or buffers were used.

Protein digests

- Typical in-solution digest ratios are 1:20 – 1:50 (enzyme weight : protein weight).
- Digestion conditions include mixing 5.0 μL of 1mg/mL denatured protein with 0.25 μL trypsin (0.1 mg/mL), 7.5 μL 200 mM NH$_4$HCO$_3$ buffer, and 6 μL H$_2$O allowing to digest at 37° C overnight.
- In preparation for MALDI analysis, deposit 0.5 μL of proteoltyic peptide solution (~1 pmol/μL) on plate and then immediately add

  0.5 μL matrix solution.

Using the dry drop method of sample crystallization, 0.1 to 1.0 picomole are introduced onto a MALDI plate, after which an equal amount of matrix solution is added to the sample. The mixture is then set to dry at room temperature or placement in a 37°C incubator can assist sample drying. Washing the dried sample/matrix with cold water can serve to remove contaminants like salts and detergents. An example of the results of washing the sample after it has been deposited on the matrix is shown in **Figure 5.5**. It is worth noting that water wash of a peptide mix can cause loss of hydrophilic peptides.

On-Plate Sample Wash

As can be seen from **Figure 5.5**, on-plate washes with cold water can be quite effective for removing salts from MALDI matrix solutions. What follows is a simple washing procedure.

- Deposit 0.5 µL of the analyte followed by 0.5 µL matrix* and dry.
- Load the sample and analyze it. Check magnified crystal appearance.
  - If results and/or crystals are of poor quality, pipette 2-3 µL of distilled water on top of the sample.
- Allow the droplet to stand for ~30 seconds to 1 minute.
  - Pipette the water droplet in and out a few times to wash the matrix solution surface.
  - Pipette the water off the sample, trying to remove as much as possible. Add a small amount of matrix, if necessary.
  - Allow to dry and analyze. The wash steps can be repeated, if necessary, but do not add matrix more than once.

*Note: we find that commercially available α-cyano matrix to typically be of low purity, a hot ethanol percepitation can enhance their performance. You will also notice that the α-cyano now is a bright yellow color instead of brown.

ZipTip™ Sample Wash

Another popular approach for cleaning samples prior to analyzing them by MALDI or electrospray is using a ZipTip™. ZipTip™ pipette tips are 10 µL (P-10) pipette tips with a bed of resin fixed at the end of each one such that there is no dead volume. ZipTip™ Pipette Tips are useful for concentrating, desalting, and fractionating picomole amounts of peptide, protein, or oligonucleotide samples prior to analysis. In operation, ZipTips are affixed to a single or multi-channel pipette. Sample is aspirated and dispensed through the tip to bind, wash, and elute the analyte(s) of interest. The concentrated, purified sample is precisely eluted in 1-4 µL of eluant. The working volume range of a ZipTip™ is between 1 to 100 µL and availability includes C_{18} and C_4

resins. The maximum molecular weight of a polypeptide which can be processed in ZipTip™ C18 is typically 50,000 Da and each tip can only be used to concentrate and desalt a sample once, which is performed in about a minute. For electrospray mass spectrometry the appropriate eluant for peptides is 50% ACN/0.1% TFA or 0.1% formic acid. For proteins 75% MeOH/0.1% TFA or 0.1% formic acid. Elution volumes for ZipTip™ Pipette Tips are typically 2-4 µL. A minimum sample amount for ZipTip™ C18 is approximately 30 femtomoles of ACTH or angiotensin. In general, 1 pmol of sample should provide an adequate signal. Recoveries of 75% are typical for concentrations above 5 pmol/µL. At concentrations of 0.25 to 2.5 pmol/µL, recoveries of 45-50% have been observed.

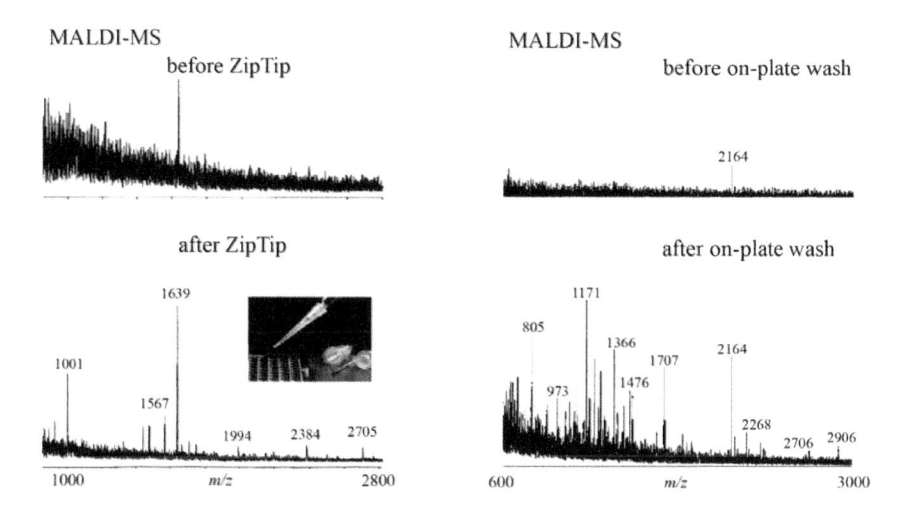

Figure 5.5. (left) MALDI-MS of a tryptic digest of 5 pmol of BSA before and after processing with C18 ZipTip. (right) MALDI-MS before and after an on-plate wash of an unknown results in the identification of mSin3A mouse protein using database searching.

Accuracy is also important in peptide and protein analysis. MALDI mass analysis (**Figures 5.2-5.7**) is typically performed with time- of-flight (TOF) on proteins or TOF-reflectron analyzers on peptides with resolving capabilities on the order of 400 to 10,000 respectively and accuracy ranging from ±0.2% to 0.001% respectively. The resolution and accuracy depend upon the type of instrument being used (e.g. low resolution TOF versus a higher resolution TOF reflectron), the presence of an internal standard, buffers, salts, contaminants, the size/type of peptide/protein, and the selection of matrix material.

MALDI-TOF of a BSA digest

54% coverage

ESI-TOF of a BSA digest

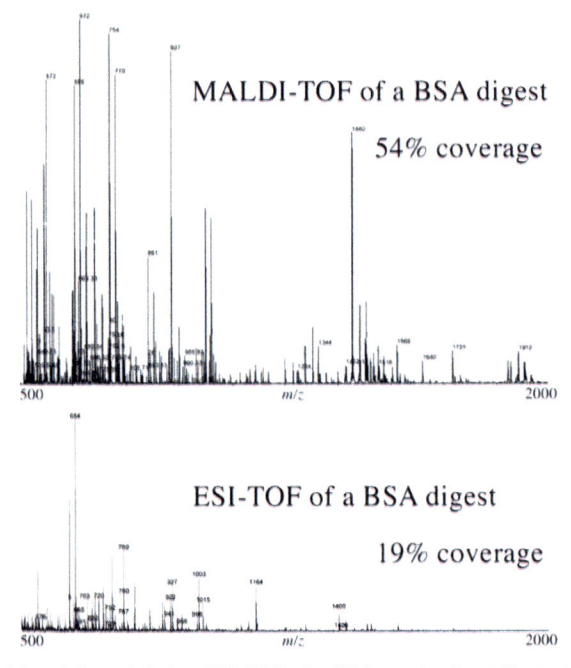

19% coverage

Figure 5.6. MALDI and direct infusion ESI-TOF of a BSA tryptic digest illustrates MALDI utility for mixture analysis. ESI mass spectral data of the same digest gave significant signal suppression of the ions. However, ESI can more easily be coupled with LC separation approaches which dramatically improves its performance for mixtures.

MALDI

viral capsid proteins

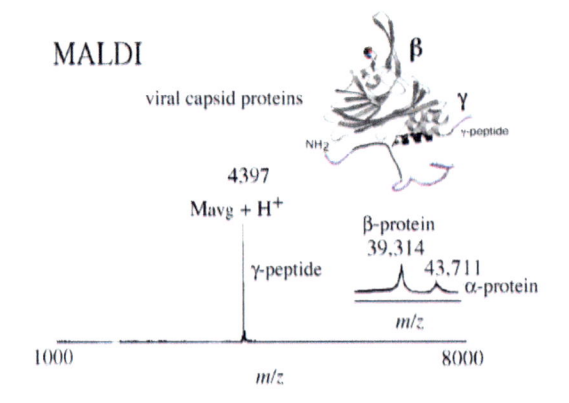

4397
$M_{avg} + H^+$

γ-peptide

β-protein
39,314
43,711
α-protein

Figure 5.7. MALDI-TOF MS data showing viral capsid proteins using sinapinic acid matrix. The data demonstrates the simultaneously acquisition of ions of significantly differing size.

Another important feature for protein/peptide analysis is mass range (**Figure 5.7**). MALDI-TOF can routinely analyze compounds from a mass of 700 to 200,000 Daltons or greater whereas MALDI-TOF reflectron is useful up to ~5000 m/z. The fact that MALDI predominantly generates ions that are singly charged makes it easy to identify individual peptides or proteins. MALDI is also the most reliable technique for analyzing glycoproteins, where the broadness in their peaks reflects the carbohydrate heterogeneity in the proteins (**Figures 5.8**).

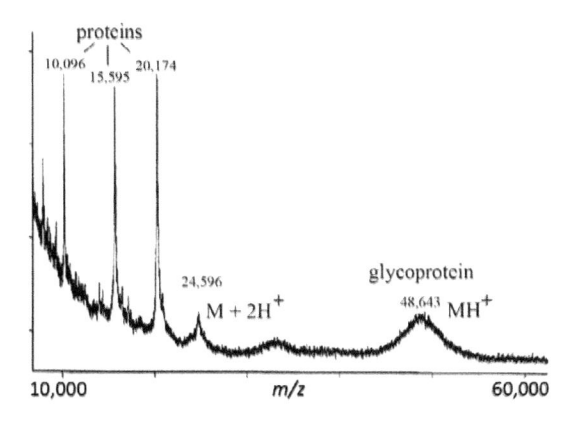

Figure 5.8. MALDI-TOF data showing a broad glycoprotein peak (m/z 48,600) and smaller proteins (no glycosylation). Glycoprotein broadness is due to carbohydrate heterogeneity.

Peptides and Protein by Electrospray

MALDI and ESI represent very different types of ionization sources for peptide and protein analysis, with each having specific advantages. Perhaps the most important of these advantages include MALDI's ability to handle complex mixtures directly while ESI best handles mixtures when interfaced with liquid chromatography. In the low mass range, MALDI matrix interference effects measurements below m/z 700 making ESI more desirable. Another consideration is that ESI is a softer ionization method and is more suitable for the analysis of noncovalent complexes. For protein analysis, ESI analysis also tends to provide better resolution and accuracy since the measurements are performed on multiply charged species in a region (below m/z 3000) where the mass analyzer has high resolving power (**Figure 5.9**).

peptide MH+ 1978
ESI - quadrupole resolution = 1500
protein MH+ 66,420

MALDI - TOF
res = 1500
res = 200

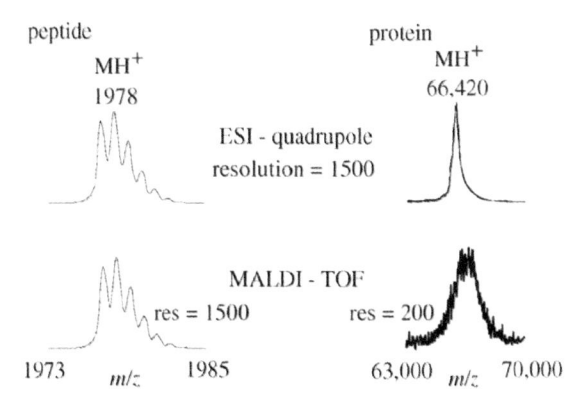

1973   m/z   1985   63,000   m/z   70,000

Figure 5.9. A direct comparison of peptide and protein data obtained from ESI-quadrupole and MALDI-TOF mass spectrometers. The data for the peptide is quite comparable, while the extra broadness of the peak associated with the protein MALDI-TOF spectrum reflects its lower resolution and lower accuracy in the higher mass range. The ESI protein data is generated from multiply charged ions below m/z 3000.

ESI is also useful for peptide and protein analysis for the following reasons:

- compatibility with liquid chromatography
- compatibility with tandem mass analyzers
- high sensitivity (subpicomole with nanoESI)
- allows for multiple charging and therefore the analysis of proteins with limited m/z range analyzers
- multiple charging allows for more complete fragmentation with collision-induced dissociation (CID).

As with MALDI, the electrospray ionization of peptides and proteins involves the addition of a proton or multiple protons. Sample preparation is achieved by dissolving the sample in a protic volatile solvent system that is relatively homogeneous and contains less than one millimolar concentration of salt, although higher concentrations have been used (such as the volatile buffer NH_4Ac at 100 mM). However, some salts (alkali and alkaline) and phosphate buffers are more detrimental to signal due to signal suppression. Because of the good mass accuracy/stability associated with quadrupole analyzers and because ESI can be easily interfaced with liquid chromatography, electrospray is often the method of choice for peptides. (Sample spectra

5. Proteomics

shown in **Figures 5.10-5.12**). Moreover, since ESI has similar sensitivity, there are only a few preparation procedures and ESI instruments tend to provide higher resolution/accuracy, it is generally the method of choice.

Multiple Charging

Another important feature of electrospray is its ability to generate multiply-charged species. Multiple charging makes it possible to observe large proteins with mass analyzers that have a relatively small *m/z* range. In addition, observing multiple peaks for the same peptide allows one to make multiple molecular weight calculations from a single spectrum. Thus, one can average these values and obtain a more accurate molecular weight. The question often arises, especially concerning the analysis of peptides and proteins, "How is the charge state of the observed ion determined?" This is accomplished from the isotope pattern or using adjacent charge states (**Figures 5.10 and 5.11**).

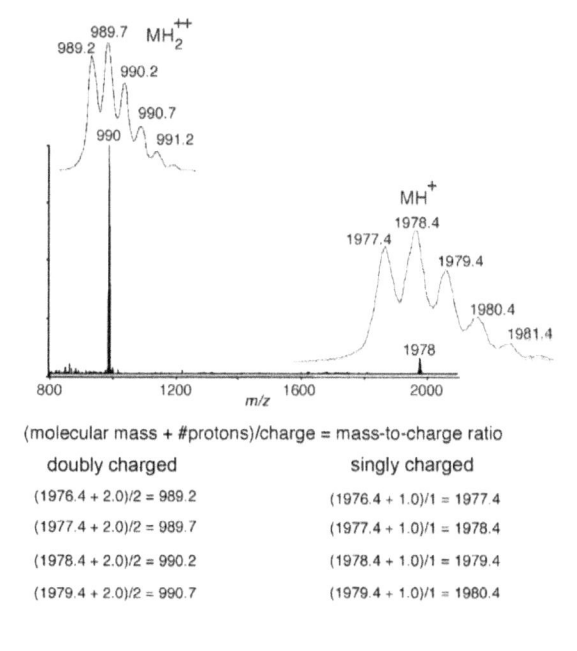

(molecular mass + #protons)/charge = mass-to-charge ratio

doubly charged	singly charged
(1976.4 + 2.0)/2 = 989.2	(1976.4 + 1.0)/1 = 1977.4
(1977.4 + 2.0)/2 = 989.7	(1977.4 + 1.0)/1 = 1978.4
(1978.4 + 2.0)/2 = 990.2	(1978.4 + 1.0)/1 = 1979.4
(1979.4 + 2.0)/2 = 990.7	(1979.4 + 1.0)/1 = 1980.4

Figure 5.10. The question of whether a peptide (or any molecule) observed is singly or multiply charged can be addressed in a several ways. The spacing between the isotopes is reliable in that if isotopes are separated by one mass unit, the charge is 1+, if by $1/2$ mass units, the charge is 2+, if by $1/3$, the charge is 3+, and so on.

Peptides and Protein by Electrospray

In the spectrum in **Figure 5.10**, two peaks are seen, one at m/z 1978 and another at m/z 990. It is often possible to look at the isotopic distribution of the ions. One mass unit separation at m/z 1978 and $1/2$ mass unit separation at m/z 990 correspond to the 1+ and 2+ charge states, respectively. The resolving power of the common quadrupole electrospray mass spectrometers only allow for distinguishing between singly and doubly-charged species.

Fortunately, as bigger peptides and proteins are analyzed, a distribution of multiply charged ions is obtained (**Figure 5.11**). Even though one cannot identify the individual charge states, from the isotopic pattern, unless using a high resolution FTMS instrument, the charge state can be deduced from distribution. **Figure 5.11** illustrates the results of two computer-generated deconvolution calculations.

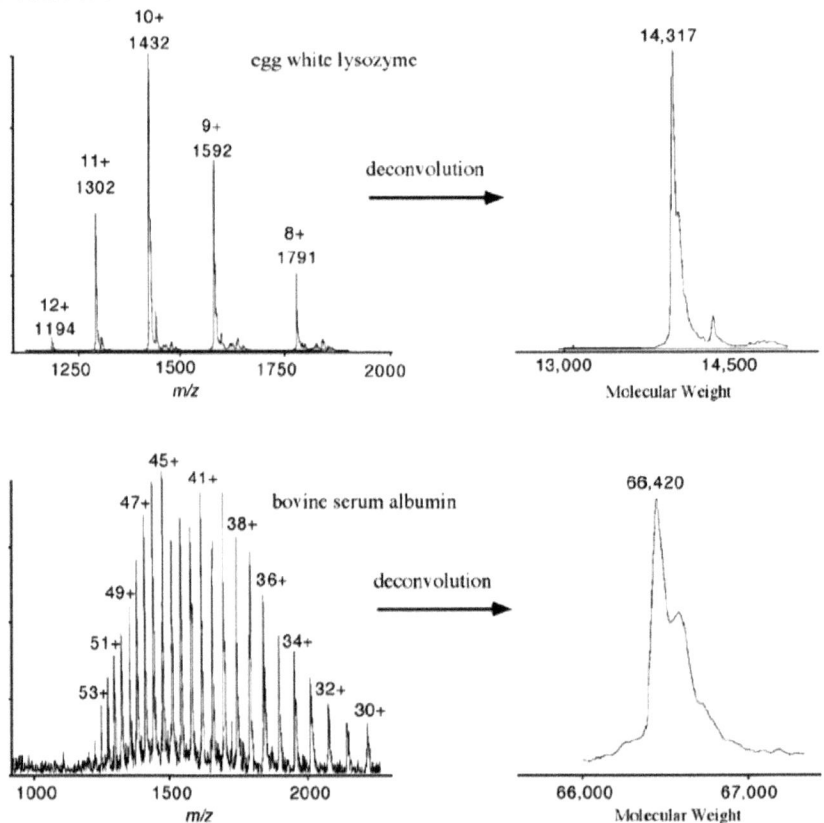

Figure 5.11. Electrospray mass spectra with the deconvoluted molecular weight spectrum of egg white lysozyme and bovine serum albumin (BSA).

5. Proteomics

Another advantage of generating multiply-charged ions with electrospray is that multiply-charged peptide ions tend to give more complete fragmentation spectra. This has become increasingly important for *de novo* sequencing of peptides and protein identification.

One of the limitations of ESI in comparison to MALDI is its lower sensitivity. However, this has been overcome with the commercial introduction of nanoelectrospray ionization sources. NanoESI is a slight variation on ESI such that the spray needle has been made very small and positioned very close to the entrance of the mass spectrometer. The end result is increased efficiency which greatly reduces the amount of sample needed. The flow rates for nanoESI are on the order of tens of nanoliters per minute and the droplets are smaller than with normal ESI. Since the droplets are smaller with nanoESI, the amount of evaporation necessary to obtain ion formation is much less. As a consequence, nanoESI is more tolerant of salts and other impurities because less evaporation means the impurities are not concentrated down as much as they are in ESI.

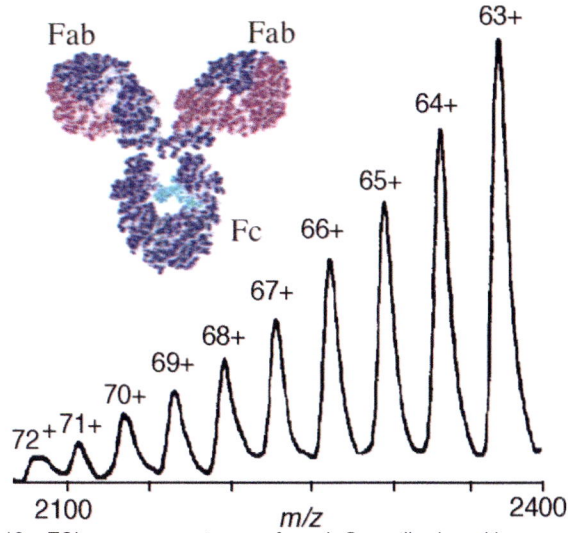

Figure 5.12. ESI mass spectrum of a IgG antibody with averaged molecular weight = 149,599 ± 12 Da. Adapted from Feng, R; Konishi, Y. "Analysis of Antibodies and Other Large Glycoproteins in the Mass Range of 150000-200000 Da by Electrospray Ionization Mass Spectrometry", *Analytical Chemistry*, 1992, 64:2090-2095.

ESI can also be applied to larger proteins. In fact, one of the largest proteins observed with ESI-MS has been with an IgG (**Figure 5.12**). However, direct analysis of these as well as glycoproteins by electrospray provides only limited information and requires very high purity because excessively heterogeneous compounds produce complicated spectra due to multiple charging, often making interpretation difficult or impossible. In addition, sample heterogeneity will often reduce instrument sensitivity via signal suppression. In general electrospray is not the method of choice and MALDI is typically used for the analysis of glycoproteins (**Figure 5.8**).

LC/MS

Another important feature of ESI-MS with regard to peptide and protein analysis is its ability to directly analyze compounds from aqueous or aqueous/organic solutions, a feature that has established the technique as a convenient mass detector for liquid chromatography (LC) (**Figure 5.13**). ESI or nanoESI combined with quadrupole or ion trap analyzers allow for MS analysis at flow rates ranging from 2mL/min to 50nL/min, respectively. The lower flow rates allow for high sensitivity measurements, such as in protein digest analysis, and the high flow rates allow for high throughput applications.

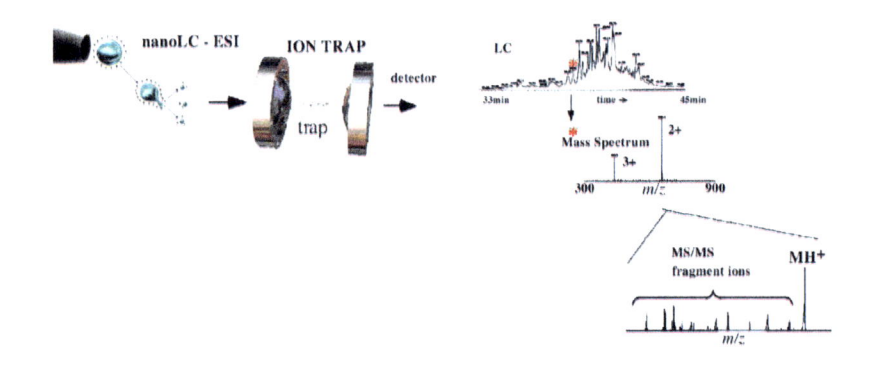

Figure 5.13. Interfacing liquid chromatography with ESI MS/MS.

5. Proteomics

Declustering

As previously described, ESI induces ion formation from the generation of small-charged droplets. Once ions are formed, they are subject to collisions as they enter the mass analyzer. These collisions can decluster aggregates, induce fragmentation, and change the charge states by removing protons. The kinetic energy of the ions can be adjusted to allow for some control over source fragmentation. This energy can be adjusted by varying the electrospray declustering or fragmentation potential. A diagram of the orifice is shown in **Figure 5.14**. However, it should be pointed out that this design is variable depending on the instrument, yet the same effect is observed in virtually all electrospray mass spectrometers. The energy of the ions entering the orifice is determined by the voltage applied to the orifice; the higher the energy, the more fragmentation is observed. In addition to inducing peptide fragmentation, increasing the declustering potential can cause the charge distribution of a protein to change. This is related to the greater collision-causing protons being stripped from the protein, and thus a shift in the charge state distribution occurs. **Figure 5.14** demonstrates that at a high declustering potential, protons are stripped from egg white lysozyme to produce a charge distribution completely different from one observed at a lower potential.

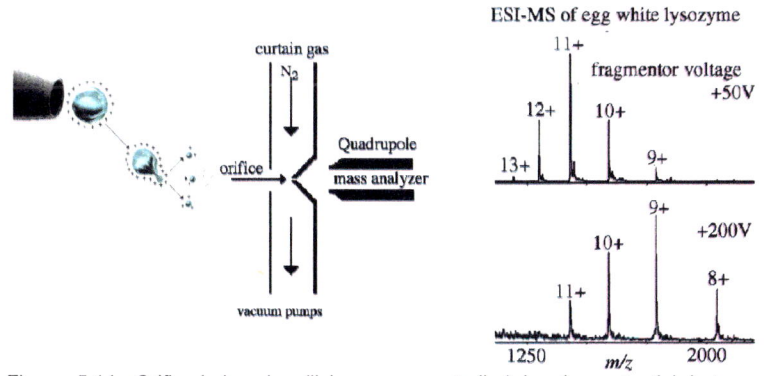

Figure 5.14. Orifice-induced collisions are controlled by the potential between the orifice and the quadrupole mass analyzer. The variation of this potential can affect sensitivity, the ability to observe noncovalent interactions, fragmenation, and charge state. The potential difference between the orifice and the quadrupole affects the energy of collisions prior to mass analysis. Electrospray mass spectra of egg white lysozyme at two different orifice (declustering) potentials demonstrates that at higher potentials, protons can be stripped from the protein thereby shifting the charge state distribution.

Another interesting aspect of the ESI source is that it can induce fragmentation of small peptides typically below a mass of 3000 Da. In relation to peptide/protein analysis, the declustering potential allows for peptides to be fragmented, thus allowing for sequencing information to be obtained. **Figure 5.15** illustrates how sequence information was obtained on a peptide simply by increasing this potential. A drawback of the application is that it can only be used on very pure samples; if the samples are not pure it may be easy to mistake an impurity for a fragment ion. When using orifice-induced fragmentation, identifying true fragment ions versus impurities is difficult. This problem can be overcome with the use of a tandem mass spectrometer. Sample purity is not as significant an issue with a tandem mass spectrometer since an ion of interest can be isolated even in the presence of several other ions (peptides). Once the ion of interest is isolated, it can be exposed to collisions and the resultant fragment ions can be mass analyzed. Tandem mass analysis and the specific applications of tandem mass spectrometry to peptides are described in the following section.

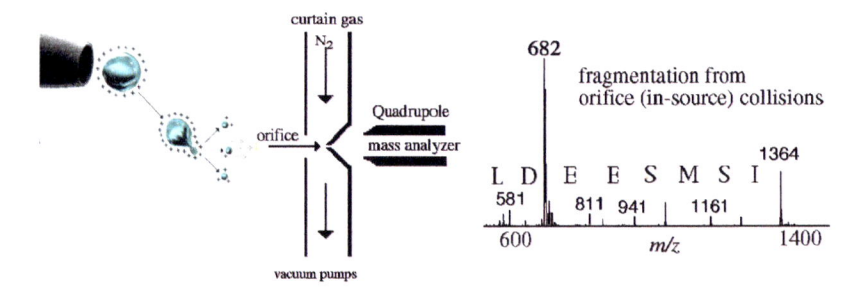

Figure 5.15. Collisions at the orifice can be useful for obtaining structural information. The collisions of this peptide, H-[ISMSEEDLLNAK]-OH, have resulted in the formation of fragment ions from the N-terminal directly corresponding to its sequence. It is important to note that this is not a tandem mass spectrometry experiment and the fragment ions were generated as the ions entered the mass analyzer.

Peptide and Protein Analysis Overview

Mass spectrometry techniques such as ESI and MALDI have changed the way biochemists are approaching the challenges of protein/peptide analysis. The availability of these sensitive and gentle

5. Proteomics

ionization methods has made it possible to analyze proteins and peptides through a range of available mass analyzers including TOFs, triple quadrupoles, quardupole ion traps, FTMS, and hybrid instruments. Furthermore, MS is now routinely used as a fundamental tool for identifying proteins.

Protein Mass Mapping

Mass spectrometry has rapidly become one of the most important tools for the identification of proteins (**Figure 5.16**) largely because of the completion of many genome sequences. Using ESI or MALDI-MS, intact proteolytic peptides can be analyzed, and their masses accurately measured. Based on this information, proteins can readily be identified using a methodology called protein mass mapping or peptide mass mapping, in which these measured masses are compared to predicted values derived from protein databases. Further sequence information can also be obtained by fragmenting individual peptides in tandem MS experiments. In addition, large scale changes in protein expression levels (protein profiling) between two different samples can be assessed using quantitative tools such as two-dimensional gel electrophoresis (2D-GE) or stable isotope labeling in conjunction with mass spectrometric measurement.

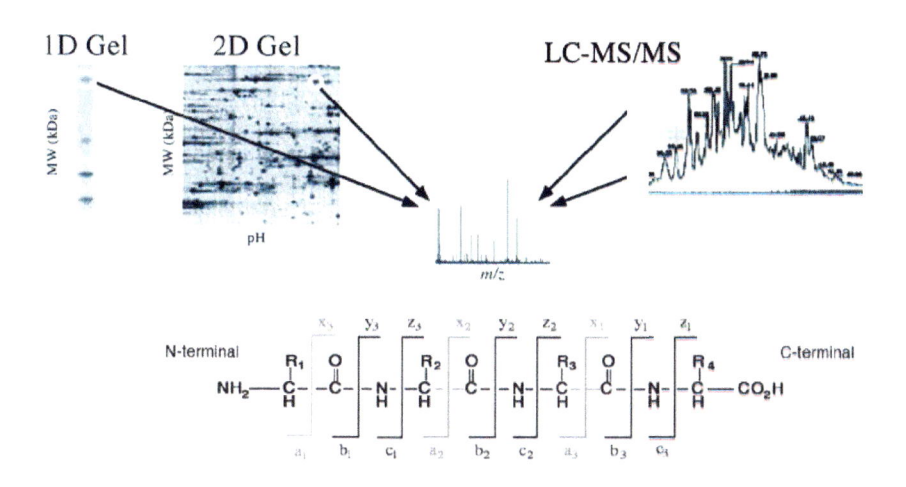

Figure 5.16. MALDI and LC/MS/MS mass spectrometry strategies used for protein ID.

Peptide Mass Mapping

Protein identification has traditionally been accomplished by subjecting proteolytic digests to high performance liquid chromatography (HPLC) followed by *N*-terminal (Edman) sequencing and/or amino acid analysis of the separated peptides. However, these techniques are relatively laborious, insensitive, and do not work with *N*-terminally modified peptides. More recently, mass spectrometry has been combined with protease digestion to enable peptide mass mapping. Definitively, peptide mass mapping (also known as protein/peptide mass fingerprint or PMF) combines enzymatic digestion, mass spectrometry, and computer-facilitated data analysis for protein identification. Sequence specific proteases or certain chemical cleaving agents (**Table 5.1**) are used to obtain a set of peptides from the target protein that are then mass analyzed. The enzyme trypsin is a commonly used protease that cleaves peptides on the *C*-terminal side of the relatively abundant amino acids arginine (Arg) and lysine (Lys). Thus, trypsin cleavage results in many reasonably sized fragments from 500 to 3000 Daltons, offering a significant probability for unambiguously identifying the target protein. The observed masses of the proteolytic fragments are compared with theoretical digests of all the proteins listed in a sequence database (**Figure 5.17**). The matches or "hits" are then statistically evaluated and ranked according to the highest probability.

Clearly, the success of this strategy is predicated on the existence of the correct protein sequence within the database searched. However, the quality and content of such databases are continually improving because of genomic sequencing of entire organisms, and the likelihood for obtaining matches is now reasonably high. While exact matches are readily identified, proteins that exhibit significant homology to the sample are also often identified with lower statistical significance. This ability to identify proteins that share homology with poorly characterized sample species makes peptide mass mapping a valuable tool in the study of protein structure and function.

Table 5.1. Protease specificity. Proteolysis experiments can use any of a number of enzymes to perform digestion. The cleavage specificity of some of the different enzymes is denoted by a slash (/) before or after the amino acid responsible for specificity. Combinations of proteases can be used to reduce specificity and to mimic other proteases. For example Lys-C and clostripain together are specific for the same sites as trypsin.

6. Proteomics

Protease	AminoAcid Specificity	Exceptions
Trypsin	X-Lys/-Y X-Arg/-Y	Does not cleave if Y = Pro
Endoproteinase Lys-C	X-Lys/-Y	Does not cleave if Y = Pro
Clostripain	X-Arg/-Y	
Endoproteinase Asp-N	X-Asp/-Y, X-cysteicacid bonds	Does not cleave if Y = Ser
CNBr	X-Met/-Y	Does not cleave if Y = Ser, Thr, or Cys
Glu-C (V8 Protease (E))	X-Glu/-Y X-Asp/-Y	Does not cleave if X = Pro
Pepsin	X-Phe/-Y X-Leu/-Y X-Glu/-Y	Does not cleave if Y = Val, Ala, Gly
Endoproteinase Arg C	X-Arg/-Y	Does not cleave if Y = Pro
Thermolysin	X-/Phe-Y X-/Ile-Y X-/Leu-Y X-/Ala-Y X-/Val-Y X-/Met-Y	Does not cleave if X = Pro
Chymotrypsin	X-Phe/-Y X-Tyr/-Y X-Trp/-Y X-Leu/-Y	Does not cleave if Y = Met, Ile, Ser, Thr, Val, His, Glu, Asp
Formic Acid	X-/Asp-Y	

Upon submitting a query to a search program, a theoretical digest of all the proteins in the database is performed according to the conditions entered by the researcher. Variables that can be controlled include taxonomic category, digestion conditions, the allowable number of missed cleavages, protein isoelectric point (pI), mass ranges, possible post translational modifications (PTMs), and peptide mass measurement tolerance. A list of theoretical peptide masses is created for each protein in the database according to the defined constraints, and these values are then compared to the measured masses (**Figure 5.17**). Each measured peptide generates a set of candidate proteins that would produce a peptide with the same mass under the digestion conditions specified. The proteins in these sets are then ranked and scored based on how closely they match the entire set of experimental data.

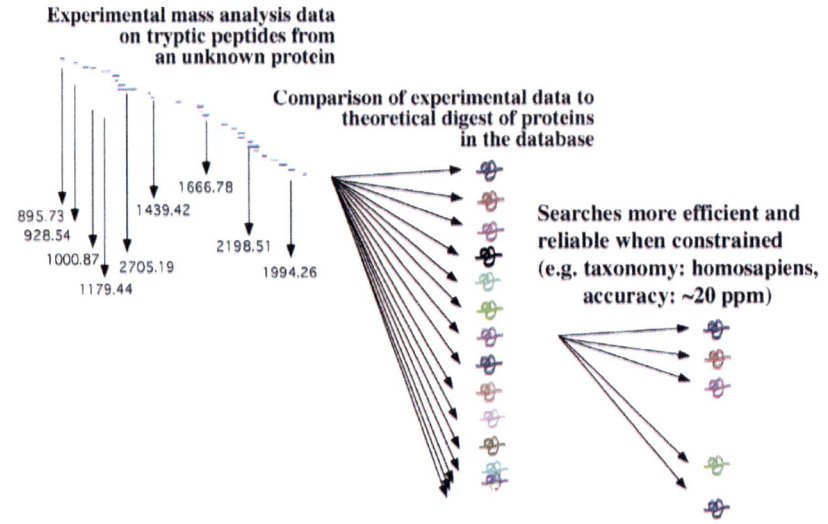

Figure 5.17. Protein identification through the comparison of tryptic peptides of an unknown protein to the theoretical digest of known proteins. The identification can be made to be more accurate when constraints are added such as the taxanomic category of the protein, as well as when high accuracy and MS/MS data are used.

This method of identification relies on the ability of mass spectrometry to measure the masses of the peptides with reasonable accuracy, with typical values ranging from roughly 5 to 50 ppm (5 ppm =
±0.005 Daltons for a 1,000 Dalton peptide). The experimentally measured masses are then compared to all the theoretically predicted peptide digests from a database containing possibly hundreds of thousands of proteins to identify the best possible matches. Various databases (**Table 5.2**) are available on the Web and can be used in conjunction with such computer search programs as Profound (developed at Rockefeller University), Spectrum Mill (originated at University of California, San Francisco) and Mascot (Matrix Science, Limited). One obvious limitation of this methodology is that two peptides having different amino acid sequences can still have the same exact mass. In practice, matching 5-8 different tryptic peptides (within 50-100 ppm accuracy) is usually sufficient to unambiguously identify a protein with an average molecular weight of 50 kDa, while a greater number of matches may be required to identify a protein of higher molecular weight. It is important to note that the term protein identification as used here

does not imply that the protein is completely characterized in terms of its entire sequence, as well as all its PTMs. Rather, this term typically refers to matching the peptides to some percentage of the amino acid sequence contained in the database. The entries in the database are both from direct protein sequence data (swissprot) and translated from genes (Genbank part of NCBI).

Table 5.2. Two of the Protein Databases Available on the Internet.

NCBInr   A largely non-redundant database compiled by the National Center for Biotechnology Information (NCBI) by combining most of the public domain databases (Expressed Sequence Tags (EST's) not included).

Swiss Prot   A curated protein sequence database which strives to provide a high level of annotations, such as the description of the function of a protein, its domain's structure, post-translational modifications, variants, etc. This database offers a minimal level of redundancy and high level of integration with other databases.

In theory, accurate mass measurements of the undigested protein could also be used for protein identification. In practice, however, the identification of a protein based solely on its intact masses is very challenging due to the stringent sample purity required, the need for extremely accurate mass measurements, and most importantly, the unpredictable variability introduced by numerous possible post- translational modifications (PTMs).

Identification Using Tandem Mass Spectrometry

A more specific database searching method involves the use of partial sequence information derived from MS/MS data (**Figure 5.18**). As discussed later, tandem mass spectrometry experiments yield fragmentation patterns for individual peptides. Manual interpretation of a tandem MS experiment can often be quite difficult due to the number of different fragmentations that can occur, not all of which yield structurally useful information. However, analogous to peptide mapping experiments, the experimentally obtained fragmentation patterns can be compared to theoretically generated MS/MS fragmentation patterns for the various proteolytic peptides arising from each protein contained in the searched database. Statistical evaluation of the results and scoring algorithms using search engines such as Sequest (ThermoFinnigan Corp) and MASCOT (Matrix Science, Limited) facilitate the identification of the best match (**Figure 5.18**). The partial sequence information contained in tandem MS experiments is much more specific than simply using the mass of a peptide, since two peptides with identical amino acid contents but different sequences will exhibit different fragmentation patterns.

Protein Characterization by Mass Spectrometry

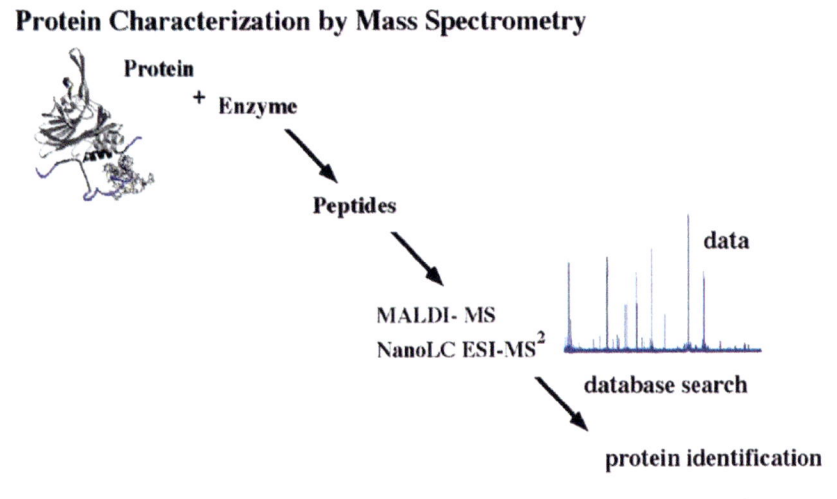

Protein + Enzyme → Peptides → MALDI-MS / NanoLC ESI-MS^2 → data → database search → protein identification

Figure 5.18. Protein identification through the comparison of tryptic peptides of an unknown protein to the theoretical digest and theoretical MS/MS data of known proteins, is even more reliable than just comparing the mass of the fragments.

While the molecular weight information obtained from ESI, and MALDI are useful in the preliminary stages of characterization, it can also be very important to gain more detailed structural information through fragmentation. Tandem mass spectrometry, the ability to induce fragmentation and perform successive mass spectrometry experiments on these ions, is generally used to obtain this structural information (abbreviated MS^n - where **n** refers to the number of generations of fragment ions being analyzed).

One of the primary processes by which fragmentation is initiated is known as collision-induced dissociation (CID). CID is accomplished by selecting an ion and then subjecting that ion to collisions with neutral atoms. The selected ion will collide with a collision gas such as argon, resulting in fragment ions which are then mass analyzed. CID can be accomplished with a variety of instruments, most commonly using triple quadrupoles, quadrupole ion traps, Fourier transform mass spectrometry (FTMS), and quadrupole time-of-flight (QTOF) mass analyzers. The quadrupole ion trap combined with electrospray ionization is currently the most common means of generating peptide fragmentation data, as they are capable of high sensitivity, and produce consistently high degrees of fragmentation information.

In order to obtain peptide sequence information by mass spectrometry, fragments of an ion must be produced that reflect structural features of the original compound. Fortunately, most peptides are linear molecules, which allow for relatively straightforward interpretation of the fragmentation data. The process is initiated by converting some of the kinetic energy from the peptide ion into vibrational energy. This is achieved by introducing the selected ion, usually an $(M + H)^+$ or $(M + nH)^{n+}$ ion, into a collision cell where it collides with neutral Ar, Xe, or He atoms, resulting in fragmentation. The fragments are then monitored via mass analysis. Tandem mass spectrometry allows for a heterogeneous solution of peptides to be analyzed and then by filtering the ion of interest into the collision cell, structural information can be derived on each peptide from a complex mixture. The fragment ions produced in this process can be separated into two classes. One class retains the charge on the N-terminus (**Figure 5.19**) and fragmentation occurs at three different positions, designated as types a_n, b_n, and c_n. The second class of fragment ions retains the charge on the C-terminus and fragmentation occurs at three different positions, types x_n, y_n, and z_n. The important thing to remember is that most fragment ions generated from ion traps, triple quadrupoles, QTOF's and FTMS instruments are obtained from cleavage between a carbonyl and a nitrogen (the amide bond). It is important to reiterate that when the charge is retained on the N-terminus this cleavage is a b-type, and when the charge is retained on the C-terminus of the peptide the cleavage is y-type.

Typical Peptide

[Chemical structure of a typical peptide showing NH₂-CH(R₁)-CO-NH-CH(R₂)-CO-NH-CH(R₃)-CO-NH-CH(R₄)-CO₂H]

Peptide Fragmentation

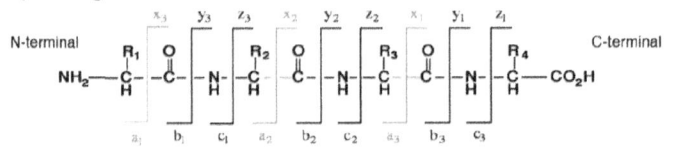

[Chemical structure showing peptide fragmentation with labeled ion positions: x₃, y₃, z₃, x₂, y₂, z₂, x₁, y₁, z₁ and a₁, b₁, c₁, a₂, b₂, c₂, a₃, b₃, c₃, from N-terminal to C-terminal]

Figure 5.19. Peptide fragmentation through collision-induced dissociation (CID) often results in the dominant fragmentation at the amide bonds in the polyamide backbone, producing ions of the type B or Y.

Certain limitations for obtaining complete sequence information exist using tandem mass spectrometry. In determining the amino acid sequence of a peptide, it is not possible for leucine and isoleucine to be distinguished because they have the same mass. The same difficulty will arise with lysine and glutamine (they differ by 0.036 Daltons) since they have the same nominal mass, although high resolution tandem analyzers (QTOF and FTMS) can distinguish between these amino acids. Another important point is that because a complete ion series (y- or b- type) is not usually observed, the combination of the two series can provide useful information for protein identification.

The Requirement for Sample Separation

Separation of proteins, or the peptides generated from a proteolytic digest, is especially important when trying to identify more than a couple of proteins simultaneously. Direct analysis by mass spectrometry in a typical biological sample is problematic due to the significant signal suppression caused by complex mixtures (**Table 5.3**). Tryptic digestion of a typical protein can result in the production of roughly fifty peptides, while miscleavages and various PTMs can give rise to many other unique species. Thus, biologically-derived samples can contain thousands to literally millions of individual peptides in the case of whole cell extracts. By comparison, even the tryptic digestion of approximately 3-5 proteins results in a peptide mixture complex enough to cause considerable signal suppression. Thus, samples of proteins (or peptides in a proteolytic digest) must be separated by gel electrophoresis or liquid chromatography prior to mass analysis.

Table 5.3. Protein Identification with MALDI and LC-MS/MS

	MALDI TOF reflectron	LC-MS/MS
Advantages	Very fast.	In addition to molecular mass data, tandem MS measurements are performed in real time.
	Widely available.	
	Easy to perform analysis.	MS/MS information adds additional level of confirmation and good identification can be obtained on two to three peptides.
	High accuracy (10-50ppm).	
	Useful for a wide range of proteins.	Multiple proteins can be analyzed simultaneously with simple reversed-phase LC run.
		Useful for PTM identification.
	Possible to reanalyze	
		High coverage of proteins (30% to 90%) depending on the protein and amount of protein.
Disadvantages	problematic for mixtures.	Computationally intensive; large database searches can take hours to days; relatively slow.
	More peptides needed for reliable identification (5-8)	Analyses performed in real time from LC eluant so reanalysis not possible
	Typically less coverage than LC-MS/MS.	

Gel Electrophoresis

Gel electrophoresis is one of the most widely used techniques for separating intact proteins. In sodium dodecyl sulfate-polyacrylamide gel electrophoresis (SDS-PAGE), sometimes called one dimensional gel electrophoresis, the proteins are treated with the denaturing detergent SDS and loaded onto a gel. As a result, each protein becomes coated with a number of negatively charged SDS molecules directly proportional to its total number of amino acids. Upon application of an electric potential across the gel, all the proteins migrate through the gel towards the anode at a rate inversely proportional to their size. The separation is typically performed with multiple proteins of known masses run alongside the proteins of interest in order to provide a size reference. Upon completion of the separation, the proteins are visualized using any of a number of different staining agents (Coomassie, Sypro Ruby, or Silver), and the individual bands are physically excised from the gel. These excised spots are subjected to destaining, reductive alkylation, in-gel digestion, peptide extraction, and finally mass analysis for protein identification (**Figure 5.20**).

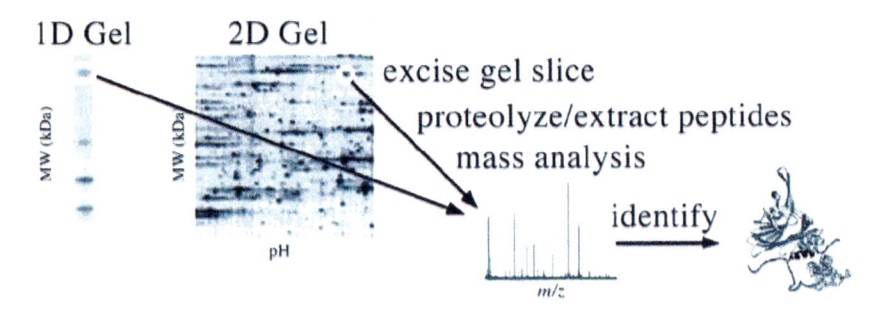

Figure 5.20. After performing SDS PAGE separation on a 1D (or 2D) gel, the stained portion of the gel representing the sample is cut out and then prepared for mass analysis by destaining, performing alkylation/reduction, in-gel digestion, and spotting on a MALDI plate. This can be done manually or robotically. Protein profiling can be performed by comparing the 2D gel from two different cell lines. The protein spot of interest is excised from the gel and an in-gel proteolysis of the protein is performed.

2-D Gels

The combination of SDS-PAGE electrophoresis with an isoelectric focusing step enables the separation of proteins of similar mass. In two-dimensional gel electrophoresis (2D-GE), proteins are first separated according to their isoelectric points (pI) by electrophoresis. In the electrophoresis step, each protein migrates to a position in the pH gradient corresponding to its isoelectric point. Once the isoelectric focusing step is complete, gel electrophoresis similar to SDS-PAGE is performed orthogonally to separate the proteins by size. Like 1D gels, 2D gel spots can be cut out, enzymatically digested, and mass analyzed for protein identification. Using this technique, thousands of proteins can simultaneously be separated and removed for identification.

In addition, 2D gels can help facilitate the analysis of certain PTMs. For example, differently phosphorylated forms of the same base protein may appear as a series of bands of roughly identical mass but different isoelectric points.

The primary application of 2D-GE is to assess large scale changes in protein expression levels between two different samples (i.e. healthy versus diseased samples). These protein profiling experiments (**Figure 5.20**) rely on the fact that the chemical stains (such as coomassie) used to visualize the separated protein bands produce responses roughly proportional to the total level of protein in the band. The experiments are typically performed by running 2D-GE on each of the samples and comparing the resulting patterns. Proteins bands that appear in only one gel or that differ significantly in their intensity are excised and identified. Alternatively, the two samples can be treated with different agents (i.e. different dyes with significantly different fluorescent emission spectra), combined, and run on the same gel.

Although still problematic, the reproducibility of 2D-GE has improved with the availability of high-quality pre-cast gels, immobilized pH gradients strips (IPG), sophisticated pattern recognition software, and laboratory automation. However, considerable limitations remain, including operational difficulty in handling certain classes of proteins, the co-migration of multiple proteins to the same position, and potential unwanted chemical modifications. An even greater potential shortcoming of the classic 2D-GE technique is its inability to accommodate the extreme range of protein expression levels inherent in complex living organisms due to sample loading restrictions imposed by the gel-based separation technology employed. It is also worth mentioning that one doesn't always know the concentrations of the samples which can result in overloading the gel and streaking. Thus, 2D- GE separations often result in only the more abundant proteins being visualized and characterized. This limitation is of particular concern in that most interesting classes of regulatory proteins are often expressed at low concentrations.

2D DIGE (Two-Dimensional Difference Gel Electrophoresis) is a technique used in proteomics for comparing and analyzing protein expression levels between different samples. Proteomics is the study of the entire set of proteins produced by a cell, tissue, or organism, and 2D DIGE is a specialized method within this field.

Here's how 2D DIGE works:

1. **Protein Extraction and Labeling**: Proteins from different samples (e.g., healthy and diseased tissues) are extracted and labeled with different fluorescent dyes. These dyes allow you to distinguish and quantify proteins from each sample separately.
2. **Isoelectric Focusing (IEF)**: In the first dimension, proteins are separated based on their isoelectric points (pI), which is a measure of their overall charge. This is done through a technique called isoelectric focusing. A pH gradient is established in a gel, and proteins migrate to positions within the gradient that correspond to their pI values.
3. **SDS-PAGE (Sodium Dodecyl Sulfate-Polyacrylamide Gel Electrophoresis)**: After the first-dimension separation, the proteins are then separated in the second dimension using SDS-PAGE. This separation is based on molecular weight, with proteins migrating through a gel matrix under the influence of an electric field.
4. **Image Analysis**: The separated proteins are visualized using fluorescent scanners. The images obtained from the gel reveal spots that represent individual proteins. The intensity of each spot corresponds to the amount of protein present in the original sample.
5. **Comparative Analysis**: The images from different samples can be compared, and software is used to identify differences in protein expression levels between the samples. Spots that show significant changes in intensity between samples are potential targets for further analysis.

2D DIGE has several advantages over traditional 2D gel electrophoresis:

- **Quantitative Comparison**: Since samples are labeled with different fluorescent dyes, 2D DIGE enables direct quantitative comparison of protein expression levels between samples.
- **Reduced Variation**: The internal standards (labeled with the same dye) help reduce gel-to-gel variation, making it easier to compare multiple gels.
- **Increased Sensitivity**: The fluorescent dyes used in 2D DIGE are often more sensitive than traditional protein staining methods, allowing for the detection of low-abundance proteins.
- **Higher Throughput**: 2D DIGE is more amenable to high-throughput studies, as multiple samples can be labeled with different dyes and run on a single gel.

5. Proteomics

Overall, 2D DIGE is a powerful tool in proteomics that enables researchers to compare protein expression profiles in a quantitative manner, leading to insights into changes in protein expression associated with various biological conditions or experimental treatments.

Table 5.4. Protein profiling with gel electrophoresis and 2D LC-MS/MS.

	2D Gels with Mass Spectrometry ID	2D LC MS/MS
Advantages	Widely available and becoming more reproducible. Software available for differential expression as well as quantifying differences (±20% at best). Easy visualization of up/down regulation demonstrated for highly complex mixtures.	Good dynamic range especially for low expression level proteins. (Codon Adaptation Index CAI of 0-0.2) very difficult for 2D gels to look below <CAI = 0.2. Good for a wide range of proteins with different hydrophobicities. MS/MS experiments performed automatically for reliable identification.
Disadvantages	problematic for proteins with extremes of PI and MW. Difficult for hydrophobic proteins. Sample handling extensive for MS analysis (stain/destain and extraction), multiple steps required and time consuming. Also increases chance of keratin contamination. This can be minimized with use of robots. Overlapping proteins possible. Limited dynamic range.	Computationally intensive; database searches can take days; cluster computer systems not readily available. Multi-dimensional protein identification technology (MudPIT) has the potential of coupling to quantitative approaches. Currently limited application shown for differential protein expression (relative quantitation) with 2D LC/MS/MS for highly complex mixtures.

Typical in-Gel Digestion Protocol for coomassie stained gels

1. Run gel and locate the protein bands of interest.
2. Excise bands of interest to 1 mm² pieces.
3. Destain by shaking for 10 minutes with 25 mM ammonium bicarbonate: 50% acetonitrile.
4. Discard – repeat until clear, rehydration of gel may be necessary.
5. Dessicate completely in speed vac.
6. Reduction: rehydrate in ~25µl 10mM dithiothreitol (DTT) or enough solution to cover the slices. Vortex, spin, and let reaction proceed 1hr at 56°C.
7. Alkylation: Remove DTT and add ~25µL 55mM iodoacetamide. Vortex, spin, and allow reaction to proceed 45 min. in the dark. Remove supernatant.
8. Wash gel slice with ~100µL ammonium bicarbonate 10 min. Discard supernatant. Wash 2x with 25mM ammonium bicarbonate/50% ACN. Speed Vac to dryness.
9. Rehydrate gel pieces in 25 mM ammonium bicarbonate pH 8.
10. Add 3µL of modified sequence grade trypsin at 0.1 µg/µL.
11. Incubate at 37°C overnight.
12. Transfer the supernatant to a new Eppendorf vial.
13. Extract the peptides using 50% acetonitrile/5% formic acid for 10 min. at 37°C.
14. Combine with the supernatant liquid from step 11.
15. Dry to approximately 10 µL and then analyze.

Adapted from Anal. Biochem. 224, 451-455 (1995) and the web site of the UCSF MS Facility.

Protein ID
MALDI-MS

The ability to profile changes in the expression levels of thousands of proteins would be relatively meaningless without the ability to rapidly identify species of interest. To this end, automated liquid handling robots have been developed that perform all the sample preparation steps for peptide mapping experiments, including gel destaining, reduction/alkylation, in gel digestion, peptide extraction, and MALDI target plating. The benefits of such automation include less potential for contamination during sample preparation, increased reproducibility, rapid protocol development, and the ability to prepare hundreds of proteins in one day. Whereas manual preparation would require a full week to perform two-hundred analyses, a robotics station can complete the task in less than one day.

Mass spectral data acquisition systems have similarly been automated to acquire spectra, process the raw data, and perform database searches for numerous samples. Commercial MALDI-TOF systems are currently available that can perform over 1,000 peptide mass mapping experiments in just twelve hours. These systems are able to perform automated calibrations, vary laser energies, and adjust laser firing location to maximize signal, with the entire data acquisition process requiring approximately 2 minutes or less. Similarly, automated data processing systems can recognize suitable signals, identify monoisotopic peaks, and submit summary peak lists directly to a search engine.

In the past, protein analyses were costly, time consuming and relatively insensitive. Today's high throughput proteomics systems enable researchers to investigate multiple unknown samples at once such as those coming from gels. Additionally, the flexibility of automated acquisition and data analysis software allows researchers to rapidly re- acquire and/or re-analyze entire batches of samples with minimal user effort.

LC-MS/MS

An alternative approach to gel electrophoresis techniques involves the use of analytical separation methods such as liquid chromatography. Although the rest of this chapter focuses specifically on liquid chromatographic techniques, it is important to note that the same advantages also apply to other separation methods such as capillary electrophoresis. Although fast and often effective for the identification of individual proteins, direct MALDI analyses usually fail when dealing with more complex mixtures due to significant signal suppression. By contrast, LC-based methodologies fractionate the peptide mixtures before MS analysis, thus decreasing signal suppression and improving the analysis of any given peptide. More importantly, additional information is obtained on individual peptides by performing tandem MS experiments.

Whereas gel electrophoresis techniques separate intact proteins, liquid chromatography can be performed on the proteolytic peptides of these proteins. One of the most popular means of performing peptide LC-MS/MS involves the direct coupling of the LC to an ion trap mass spectrometer through an electrospray ionization interface. Other mass analyzers suitable for these experiments include triple quadrupoles and QTOF's. However, ion traps and orbitraps remain the popular because of their rapid scanning capability that enables tandem mass measurements to be performed in real time. For example, the ion trap first performs MS measurements on all the intact peptide ions. Then, in a second scan, it performs an MS/MS experiment on a particular peptide. This series of alternating scans can rapidly be repeated, with different ions selected for each tandem MS experiment. In this manner, single peptides from a complex mixture can individually be addressed, fragmented, and analyzed.

The additional sequence information provided by tandem MS in the LC/MS experiments can be extremely powerful, sometimes enabling a definitive protein identification to be made based on a single peptide. Obviously, tandem MS spectra of multiple peptides that arise from the digestion of a given protein provide greater opportunity for obtaining a definitive identification. Generally, fragmentation information can be obtained for peptides with molecular masses up to 2500 Daltons. Larger peptides can reveal at least partial sequence information that often suffices to solve a particular problem.

Although complete ion series (y-type or b-type) are usually not observed, the combination of the two series often provides useful information and sometimes the entire sequence. In addition, some amino acids as well as certain PTMs bias the fragmentation towards certain cleavages (loss of phosphate), dramatically decreasing the amount of sequence information obtained. Although chemical labeling techniques can partly compensate for these phenomena, it is important to note that not every peptide yields useful tandem MS spectra, thus further emphasizing the usefulness of attempting tandem MS on multiple peptides arising from a given protein.

2D LC-MS/MS

LC-MS/MS methodologies for protein identification have been extended to mixtures of even greater complexity by performing multidimensional chromatographic separations before MS analysis (**Figure 5.21**). As its name suggests, extremely complex tryptic digests are first separated into a number of fractions using one mode of chromatography, and each of these fractions is then further separated using a different chromatographic method. In theory, any combination of operationally compatible chromatographic methods possessing sufficiently orthogonal modes of separation can be utilized, and several different combinations have been described in the literature. However, the overwhelming number of studies to date have combined strong cation exchange (SCX) and reversed-phase (RP) chromatographies. More recently, further improvements have been realized by having both chromatographic beds in a single capillary column and directly coupling this column to an ion trap mass spectrometer. A step gradient of salt concentrations is used to elute different peptide fractions from the SCX resin onto the RP material, after which RP chromatography is performed without affecting the other peptides still bound to the SCX resin. The resulting nano-RP LC column effluent is directly electrosprayed into the mass spectrometer, making this method not only amenable to automation but also very sensitive. Using this "MudPIT" methodology (Multi-Dimensional Protein Identification Technology), thousands of unique proteins have been identified from a whole cell lysate in a single 2D LC MS/MS experiment. Additionally, recent studies have also indicated that this technique possesses a greater dynamic range than that obtained using 2D gel electrophoresis, enabling the detection of lower abundance proteins. However, one limitation slowing this methodology's wide scale implementation is the computing power required to effectively compare tens of thousands of of tandem MS spectra experimentally generated. Fortunately improvements in the data analysis algorithms are gradually solving this problem.

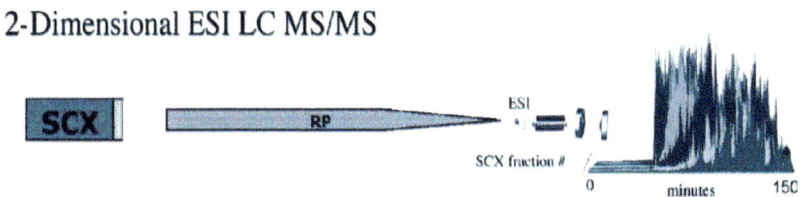

2-Dimensional ESI LC MS/MS

Figure 5.21. The proteolytic peptides separated by liquid chromatography and 2-D liquid chromatography are ionized using electrospray ionization and then subjected to tandem mass spectrometry (MS/MS) experiments. The data from the 1-D experiments can analyze up to 200 proteins simultaneously, while the 2-D experiments are capable of handling thousands of proteins.

Protein Mass Mapping and Isotope Labeling

Protein mass mapping studies can also be performed using multi- dimensional LC-MS/MS in conjunction with stable isotope labeling methodologies. Specifically, two samples to be compared are individually labeled with different forms of a stable isotopic pair, their tryptic digests combined before the final LC-MS analysis. This ideally results in every peptide existing as a pair of isotopically labeled species that are identical in all respects except for their masses. Thus, each isotopically labeled peptide effectively serves as its partner's internal standard, and the ratio of the relative heights of two isotopically labeled species provides quantitative data as to any differential change that occurred in the expression of the protein from which the peptide arose.

One approach towards differential labeling involves growing cells in isotopically enriched media. For example, one set of cells would be cultured in media that contained ^{14}N as the only source of nitrogen atoms, while the comparative case would be grown in media that only contained ^{15}N. Although effective in incorporating different stable isotopes into the two samples, the determination of which two peptides comprise an isotopically labeled pair is complicated by the fact that each pair exhibits a distinct mass difference depending on the number of isotopic atoms incorporated.

Alternatively, isotope-coded affinity tags (ICAT) (**Figure 5.22**) provides a more generally applicable approach based on the *in vitro* chemical labeling of protein samples. Specifically, ICAT utilizes the high specificity of the reaction between thiol groups and iodoacetyls (such as iodoacetamide) to chemically label cysteine residues in proteins with isotopically light or heavy versions of a molecule that differ only by the existence of eight hydrogen or deuterium atoms, respectively. The labeled protein samples are then combined and simultaneously digested, resulting in every cysteine-containing peptide existing as an isotopically labeled pair differing in mass by eight Daltons per cysteine residue. It should be noted that the general strategy of chemical labeling can be extended to other functional groups present in proteins for which chemical selective reactions exist, and several such approaches have been reported.

5. Proteomics

Protein Profiling with LC MS/MS and Isotope Labeling

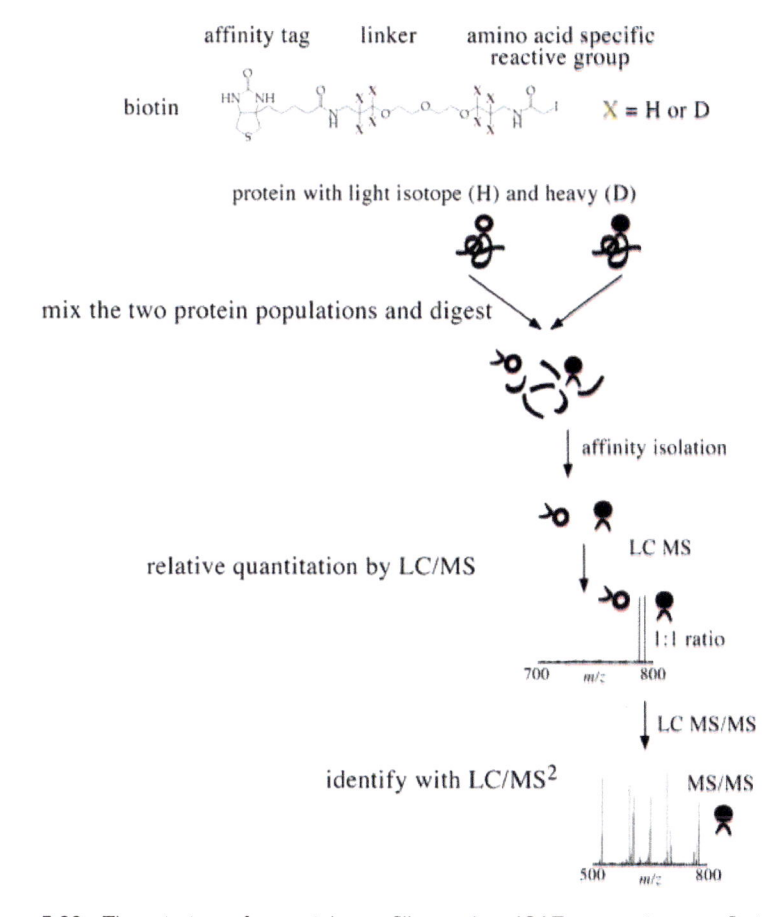

Figure 5.22. The strategy for protein profiling using ICAT reagents was first proposed by Gygi et al. (*Nature Biotechnology*) and can be broken down into 5 parts. The reagents come in two forms, heavy (deuterium labeled) and light (hydrogen labeled). Two protein mixtures representing two different cell states treated with the isotopically light (open circles) and heavy (filled circles) ICAT reagents. An ICAT reagent is covalently attached to each cysteinyl residue in every protein. The protein mixtures are combined, proteolyzed and the ICAT-labeled peptides are isolated with the biotin tag. LC/MS reveals ICAT-labeled peptides because they differ by 8 Da. The relative ratios of the proteins from the two cell states are determined from the peptide intensity ratios. Tandem mass spectrometry data is used concurrently to obtain sequence information and identify the protein. A more recent version of ICAT containing C13 labeling allows for better co- elution of the unlabeled and labeled peptides.

Due to the low natural abundance of cysteine compared to other amino acids, the overwhelming majority of the tryptic peptides remain unlabeled, and can interfere with the accurate determination of which two peptides comprise an isotopically labeled pair. Therefore, before the final LC-MS/MS analysis, an affinity selection is performed to selectively isolate the cysteine-containing species from the remainder of the tryptic peptides. In its original embodiment, a biotin molecule was also incorporated into the chemical label and used in conjunction with a monomeric avidin column to affinity purify the cysteine-containing peptides. Alternative approaches such as acid cleavable biotin are making ICAT easier to use. Although these solutions enable the accurate identification of isotopically labeled peptide pairs, they obviously preclude the analysis of proteins that do not contain cysteines and also greatly reduce the number of opportunities to effect LC-MS/MS identification of the proteins that do contain cysteine residues.

LC-MALDI MS/MS

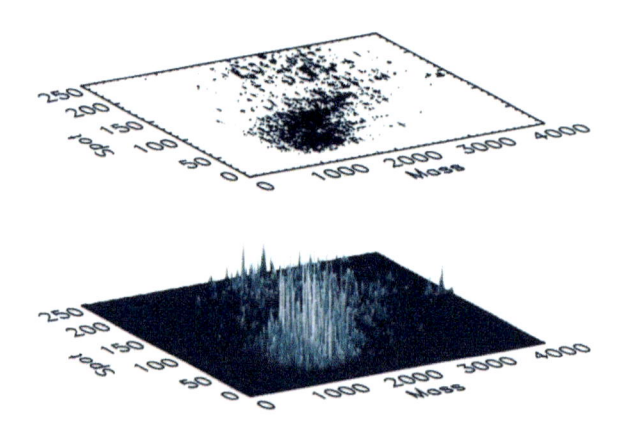

Figure 5.23. An example of HPLC MALDI-MS. (Top) Four µHPLC columns performing parallel deposition on 384 microtiter plate formats for analysis by MALDI. (Bottom) Three- dimensional plot of the reversed-phase µHPLC–MALDI FT-ICR MS analysis of a tryptic digest of yeast proteins. Image courtesy of Eric Peters *et al.*.

LC-MALDI-MS/MS strategies involve multidimensional separation by depositing the effluents of the final separation columns directly onto MALDI target plates (**Figure 5.23**). De-coupling the separation step from the mass spectrometer in this manner enables more thorough analyses of samples to be performed due to the removal of artificially imposed time restrictions. The resulting plates can also be reanalyzed as required without the need to repeat the separation step, thus decreasing sample requirements while focusing system resources only on the acquisition of tandem MS spectra of species of interest. An advantage of this approach is high throughput while the two main disadvantages are relatively poor MS/MS data from singly charged species typically generated from MALDI and quantitative analyses of protein content with MALDI is not as accurate as ESI based approaches.

Protein Mass Mapping Overview

Both MALDI and LC-MS/MS are playing important roles in protein identification and protein profiling. MALDI offers many advantages in terms of speed and ease of use, whereas LC-MS/MS offers a more reliable protein ID as well as a greater potential for quantitative analysis and post-translational modification identification. LC-MS/MS currently appears to be the primary tool for profiling.

Protein Stucture

Mass spectrometry is now commonly used for determining both the primary and higher order structures of proteins. The basis for these investigations lies in the ability of mass analysis techniques to detect changes in protein conformation under differing conditions. These experiments can be conducted on proteins alone such as monitoring charge state or in combination with proteolytic digestion or chemical modification. Proteases and chemical modification have long been used as probes of higher order structure, an approach that has been rejuvenated with the emergence of high sensitivity and accurate mass analysis techniques. Here, the applications of proteases and chemical modification with mass spectrometry are illustrated. For example, protein mass maps have been used to probe the structure of a protein/protein complex in solution (cell cycle regulatory proteins, p21 and Cdk2). This approach was also used to study the protein/protein complexes that comprise viral capsids, including those of the common cold virus where, in addition to structural information, protein mass mapping revealed mobile features of the viral proteins. Protein mass mapping clearly has broad utility in protein identification and profiling, yet its accuracy and sensitivity also allow for further exploration of protein structure and even structural dynamics.

Protein mass mapping combines enzymatic digestion, mass spectrometry, and computer-facilitated data analysis to produce and examine proteolytic fragments. This is done for the purpose of identifying proteins and, more recently, for obtaining information regarding protein structure. For protein identification, sequence-specific proteases are incubated with the protein of interest and mass analysis is performed on the resulting peptides. The peptides and their fragmentation patterns are then compared with the theoretical peptides predicted for all proteins within a database and matches are statistically evaluated. Similarly, the higher order structure of a protein can be evaluated when mass mapping techniques are combined with limited proteolytic digestion. Limited proteolysis refers to the exposure of a protein or complex to digestion conditions that last for a brief period; this is performed to gain information on the parts of the protein exposed to the surface.

The sequence specificity of the proteolytic enzyme plays a major role in the application of mass spectrometry to protein structure. A sequence-specific protease reduces the number of fragments that are produced and, concomitantly, not only improves the likelihood for statistically significant matches between observed and predicted fragment masses, but also reduces the opportunities for spurious matches. Another factor, the accessibility/flexibility of the site to the protease, also plays an important role in the analysis of structure. In this instance, ideally, only a subset of all possible cleavages is observed owing to the inaccessibility and/or inflexibility of some sites due to higher order protein structure. An example illustrated in **Figure 5.24** provides asterisks to mark potential cleavage sites that are exposed on the surface of a hypothetical protein. Since amino acids with hydrophilic side chains are found in greater abundance on the surface of proteins (at the solvent interface), proteases that cleave at hydrophilic sites are preferred in structural analysis. Trypsin and V8 protease, which cleave basic (K, R) and acidic sites (D, E), respectively, are good choices. Reaction conditions must be controlled to produce only limited proteolysis so that the cleavage pattern reliably reports on native tertiary structure. The cleavage of a single peptide bond can destabilize protein structure, causing local structural changes or even global unfolding. Therefore, subsequent protease cleavage reactions would not be useful.

5. Proteomics

theoretical proteolysis of a single protein

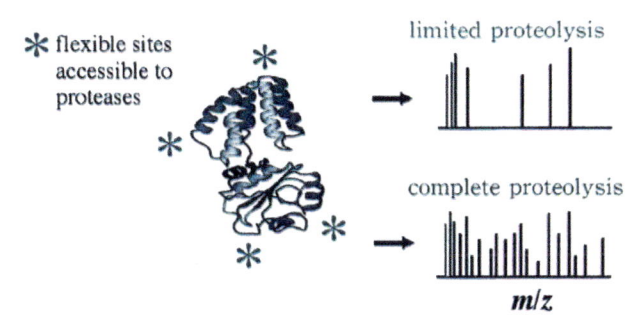

Figure 5.24. Illustration of the use of limited proteolytic cleavage as a probe of protein structure. The asterisks mark surface accessible loops that would be susceptible to proteolytic cleavage. Mass analysis of the limited proteolytic fragments yields a cleavage 'map' that provides information on structure. Complete digestion provides no structural information but is useful for protein identification.

Protein mass mapping can also be used to probe the quaternary structure of multicomponent assemblies such as protein-protein complexes, protein-DNA complexes and even intact viruses. The first application of limited proteolysis and MALDI mass spectrometry to the study of a multicomponent biomolecular assembly was performed by Chait and co-workers in 1995. In their studies, this combined approach was used to analyze the structure of the protein transcription factor, Max, free in solution as well as bound to an oligonucleotide containing its specific DNA binding site. The common feature when analyzing either protein-protein complexes or protein-DNA complexes is that the protease provides a contrast between the associated and unassociated states of the system. The formation of an interface between a protein and another macromolecule will protect otherwise accessible sites from protease cleavage and therefore provide information about residues that form the interface.

Recognizing Conformational Changes with Mass Mapping

Protein mass mapping can be used to recognize simple conformational differences between different protein states. For example, X-ray crystallography data has shown that the protein calmodulin (CaM) undergoes conformational changes in the presence of calcium. The tertiary structure of calmodulin consists of an overall dumbbell shape (148 amino acids) with two globular domains separated by a single, long central alpha-helix connector (amino acids 65-92). It has been proposed that calcium-binding activates calmodulin by exposing hydrophobic residues near the two ends of the central helix. Mass maps resulting from digests by trypsin, chymotrypsin, and pepsin all demonstrated that the protein had undergone a tertiary structural change in the presence of calcium. **Figure 5.25** shows the trypsin digests of calmodulin in the presence and absence of calcium. Comparison of the two mass spectra reveals differences corresponding to cleavages in the central helical region of the protein. Based on the results of this relatively simple experiment, it can be appreciated that structural changes caused by Ca^{2+} binding within the dumbbell domains are propagated to the central helix, as manifested by altered protease reactivity within this latter structural feature.

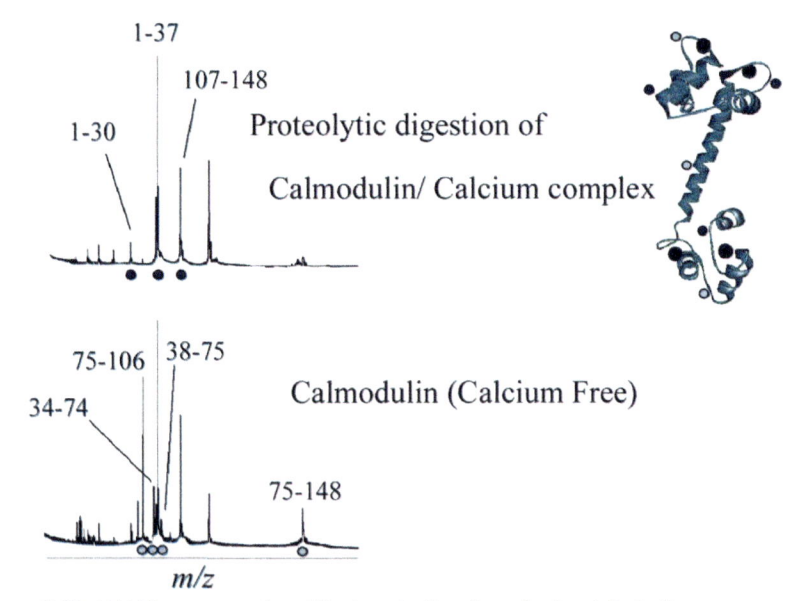

Figure 5.25. MALDI mass spectra of the trypsin digestion of calmodulin in the presence and absence of calcium. Differences are observed corresponding to cleavages within the central helical region of calmodulin (black dots) which are not observed in the presence of calcium (yellow dots) indicating a tertiary conformational change. Cleavage sites that are present in the Ca^{2+}/Calmodulin complex (black dots) and those that disappear upon addition of Ca^{2+} (yellow dots) are shown in both the spectra and the ribbon drawing. (Courtesy of J. Kathleen Lewis: Thesis, Arizona State University)

5. Proteomics

Electrospray ionization mass spectrometry has also been used to monitor protein folding and protein complexes. Early in the use of electrospray it was recognized that some proteins exhibit a distinct difference in their charge state distribution which was a reflection of their solution conformation(s). For example, two charge state distributions are shown in **Figure 5.26,** which depicts a protein's less-charged native form and the more highly charged denatured conformation. The difference between the spectra is due to the additional protonation sites available in the denatured form. This phenomenon is demonstrated for the protein fibronectin where the charge distribution is maximized at the lower charge states at +6, and the denatured protein has a distribution maximized at a higher charge state of +10. There is growing recognition of the view that proteins function through a diverse range of structural states, from highly ordered globular structures to highly flexible, extended conformations. ESI-MS is a simple but highly sensitive and informative method to characterize the functional shape(s) of proteins (*i.e.* globular or extended) prior to more material-intensive and time-consuming spectroscopic or crystallographic studies.

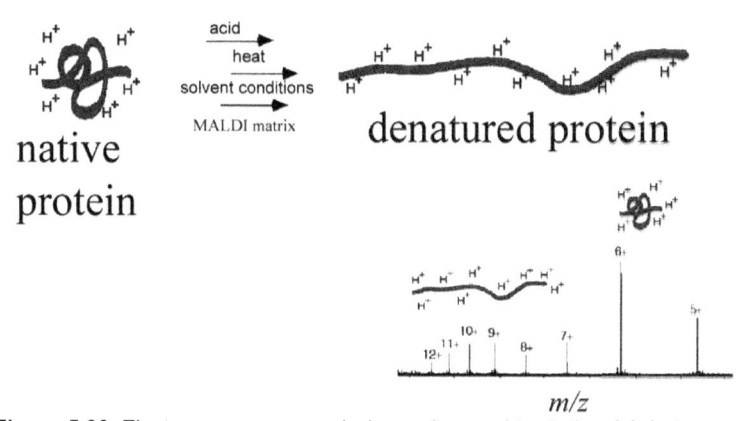

Figure 5.26. Electrospray mass analysis can be used to distinguish between native and denatured conformers of a protein. Denaturing a protein can often enhance ionization by dramatically increasing the number of sites available for protonation. The data shown represent both the native and denatured conformers of the fibronectin module. Adapted from Muir *et al.* (1995).

The use of hydrogen/deuterium (H/D) exchange to study conformational changes (**Figure 5.27**) in proteins or protein/protein interactions has been primarily performed using ESI-MS, although some studies have employed MALDI-MS. The concept of this approach is relatively simple in that amide protons within the portion of the proteins in close inter- or intra-molecular contact may form hydrogen bonds and will have different exchange rates relative to other more accessible regions of the complex. By monitoring this amide-hydrogen exchange, information can be gained on the noncovalent structure of a protein by itself or in a protein complex. It should be noted that while it is not possible through ESI-MS to monitor exchange in a residue-specific manner, populations of protein molecules with distinct masses can be distinguished. The combined application of ESI-MS with NMR spectroscopy to monitor protein folding reactions using H/D exchange is particularly powerful. For example, Dobson, Robinson, and co-workers characterized in detail transient protein folding intermediates for the protein egg white lysozyme using this dual approach. This was the first demonstration of discreet yet transient intermediates during a protein folding reaction. Importantly, these key findings were made possible only by complementing the more traditional NMR-H/D exchange approach with data from ESI-MS.

H-D exchange

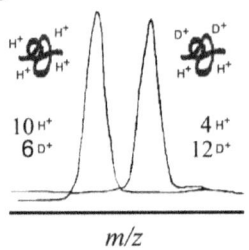

Figure 5.27. Theoretical mass spectra of two different populations of proteins – for native and denatured, where the native would be less susceptible to exchange.

Kriwacki *et al.* demonstrated the utility of limited proteolysis on the cell-cycle regulatory proteins, Cdk2 and p21-B (**Figure 5.28**). Given the known protein sequence and the sequence specificity of trypsin, the mass measurement readily identifies the exact proteolytic site within each individual protein's sequence. The results revealed a segment of 24 amino acids in p21-B that is protected from trypsin cleavage, thus identifying the segment as the Cdk2 binding site on p21-B. The concepts illustrated in this simple example can be extended to much more complex systems, allowing insights into tertiary and quaternary structure to be obtained using picomole quantities of protein.

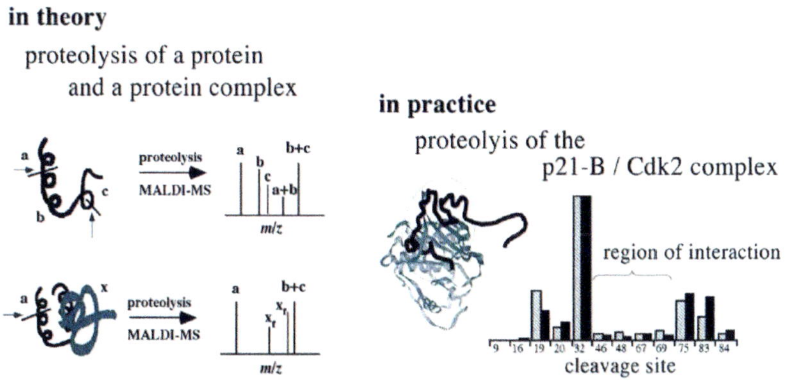

Figure 5.28. Probing protein/protein interactions using proteolysis and MALDI-MS. Schematic view (left) of key concepts. Two cleavage sites are accessible for the protein of interest (top), yielding five fragments after limited digestion. In the complex with protein X, one site is protected (bottom), yielding fewer fragments. However, fragments from protein X are also produced (X_f). Results represented as a histogram on the right indicate the "region of interaction" of p21-B with Cdk2 in a 24-amino acid segment (Kriwacki et al.).

Protein Structure Overview

The utility of combining proteolysis as well as chemical modification methods with MS analysis is becoming useful in structural studies of proteins. It is important to note that the key concepts of the methods are straightforward, and the probing reactions are simple to perform. In the early stages of structural studies, the MS-based probing methods are particularly well-suited to provide rapid access to low- resolution maps that can then be used to guide subsequent high- resolution crystallographic or NMR studies. This stage may also be an end-point in some investigations where the simple identification of interacting residues is the desired information. Probing studies have also been shown to provide insight into structural dynamics.

Questions

- What are some of the advantages of MALDI?

MALDI is sensititve and tolerant of mixtures, however its inability to effectively couple it to chromatography presents a problem.

- What is the purpose of the MALDI on-plate sample wash?

Washing helps removes salts and other contaminents that can cause signal suppression.

- What challenge does glycoprotein analysis present?

Very heterogeneous and therefore broad peak shapes.

- Why does ESI generally generate multiply-charged species of peptides and proteins?

It is such a soft ionization approach and does not strip off protons.

- Why does ESI not generally generate multiply-charged species of small molecules?

Localization of multiple charges on a small molecule will destabilize it, however on peptides and proteins the charges are sufficiently separated.

- What affect does salts and buffers have on ESI signal?

Salts and buffers alter the surface tension on the ESI droplets making it difficult or impossible to reach the charge densities required to eject ions.

- What is the advantage of ESI that allows it to be easily coupled with liquid chromatography.

ESI ions are generated directly from solutions that are very similar to those used in liquid chromatography.

- What is peptide mass mapping?

Peptide mass mapping is the ability to take proteolytic peptides and translate them into protein identifications and/or structural information.

- Why is trypsin commonly used as a protease in peptide mapping?

Trypsin is very reliable at cleaving specific amino acids.

- How is tandem mass spectrometry useful in protein identification?

MS/MS of tryptic peptides provides higher level ID confirmation.

- Why is separation of the proteins/peptides important for mass spectrometry studies?
  Separation by chromatography helps minimize signal suppression.

- How can limited proteolysis be used for protein structure?
  Limited proteolysis is useful in identifying the exposed surface of protein structure, thus giving useful insights into its solution phase conformation.

Useful References

Jonscher KR & Yates JR III. *The quadrupole ion trap mass spectrometer-A small solution to a big challenge*. Anal. Biochem. **1997**, *244*, 1-15.

Cole RB (Editor). *Electrospray Ionization Mass Spectrometry: Fundamentals, Instrumentation, and Applications*, John Wiley and Sons: New York, **1997**.

Jensen ON, Podtelejnikov A, Mann M. *Delayed extraction improves specificity in database searches by matrix-assisted laser desorption/ionization peptide maps*. Rapid Comm. Mass Spectrom. **1996**, 10, 889-896.

March RE & Todd JFJ (Editor). Practical Aspects of Ion Trap Mass Spectrometry: Fundamentals of Ion Trap Mass Spectrometry. Volume I. CRC Press: Boca Raton, **1995**.

Papayannopoulos IA. *The Interpretation of Collision-Induced Dissociation Tandem Mass Spectra*. Mass Spectrometry Reviews. **1995**, 14, 49-73.

Yates 3rd JR. *Mass spectrometry: From genomics to proteomics*. Trends in Genetics. **2000**, 16, 5-8.

Aebersold R & Goodlett DR. *Mass spectrometry in proteomics*. Chem Rev. **2001**, 101, 269-295.

Voss T & Haberl P. *Observations on the reproducibility and matching efficiency of two-dimensional electrophoresis gels: consequences for comprehensive data analysis*. Electrophoresis. **2000**, 21, 3345-3350.

Gygi SP, Corthals GL, Zhang Y, Rochon Y, Aebersold R. *Evaluation of two-dimensional gel electrophoresis-based proteome analysis technology*. Proc. Natl. Acad. Sci. USA. **2000**, 97, 9390-9395.

Corthals GL, Wasinger VC, Hochstrasser DF, Sanchez JC. *The dynamic range of protein expression: a challenge for proteomic research*. Electrophoresis. **2000**, 21, 1104-1115.

Wall DB, et al. *Isoelectric focusing nonporous RP HPLC: a two-dimensional liquid-phase separation method for mapping of cellular proteins with identification using MALDI-TOF mass spectrometry.* Anal. Chem. **2000**, 72, 1099-1111.

Washburn MP, Ulaszek R, Deciu C, Schieltz DM, Yates 3rd JR. *Analysis of quantitative proteomic data generated via multidimensional protein identification technology.* Anal. Chem. **2002**, 74, 1650-7.

Gygi SP, et al. *Quantitative analysis of complex protein mixtures using isotope-coded affinity tags.* Nature Biotechnology. **1999**, 17, 994-999.

Peters EC, Horn DM, Tully DC, Brock A. *A novel multifunctional labeling reagent for enhanced protein characterization with mass spectrometry.* Rapid. Commun. Mass Spectrom. **2001**, 15, 2387-2392.

Peters EC, Brock A, Horn DM, Phung QT, Ericson C, Salomon AR, Ficarro SB, Brill LM. *An Automated LC –MALDI FT-ICR MS Platform for High-Throughput Proteomics*. LCGC Europe. **July 2002**, 2-7.

Jonsson AP. *Mass spectrometry for protein and peptide characterisation.* Cell Mol Life Sci. **2001**, 58, 868-884.

Kriwacki RW, Wu J, Siuzdak G, Wright PE. *Probing Protein/Protein Interactions with Mass Spectromety and Isotopic Labeling: Analysis of the p21/Cdk2 Complex*. J. Amer. Chem. Soc. **1996**, 118, 5320.

Kriwacki RW, Wu J, Tennant T, Wright PE, Siuzdak G. *Probing protein structure using biochemical and biophysical methods. Proteolysis, matrix-assisted laser desorption/ionization mass spectrometry, high- performance liquid chromatography and size-exclusion chromatography of p21Waf1/Cip1/Sdi1*. Journal of Chromatography. **1997**, 777, 23-30.

Cohen SL, Ferre-D'Amare AR, Burley SK, Chait BT. *Probing the solution structure of the DNA-binding protein Max by a combination of proteolysis and mass spectrometry*. Protein Science. **1995**, 4, 1088-1099.

Fontana A, et. al. *Probing the Conformational State of Apomyoglobin by Limited Proteolysis*. J. Mol. Bio. **1997**, 266 (2), 223-230.

5. Proteomics

Fontana A, de Laureto PP, De Filippis V, Scaramella E, Zambonin M M. *Probing the partly folded states of proteins by limited proteolysis.* Folding and Design. **1997**, 2 R17-26.

Finn BE, Evenas J, Drakenberg T, Waltho JP, Thulin E, Forsen S. *Calcium-induced structural changes and domain autonomy in calmodulin.* Nat. Struct. Biol. **Sep 1995**, 2:9, 777-83.

Lewis JK. *Protein Structural Characterization Using Bioreactive Mass Spectrometer Probes.* Arizona State University Ph.D. Thesis Dissertation. **1997**.

Muir TW, Williams MJ, Kent SB. *Detection of synthetic protein isomers and conformers by electrospray mass spectrometry.* Anal Biochem. **Jan 1995**, 1:224, 100-9.

Wang L, Pan H, Smith DL. *Hydrogen exchange-mass spectrometry: optimization of digestion conditions.* Mol Cell Proteomics. **Feb 2002**, 1:2, 132-138.

Miranker A, Robinson CV, Radford SE, Aplin RT, Dobson CM. *Detection of transient protein folding populations by mass spectrometry.* Science. **1993**, 262, 896-900.

Ho Y, et. al. *Systematic identification of protein complexes in Saccharomyces cerevisiae by mass spectrometry.* Nature. **2002**, 415, 180-183

Broo KM, Wei J, Marshall D, Brown F, Smith TJ, Johnson JE, Schneemann A, Siuzdak G. *Viral Capsid Mobility: A Dynamic Conduit for Inactivation.* PNAS. **2001**, 98, 2274-2277.

Glocker MO, Borchers C, Fiedler W, Suckau D, Przybylski M. *Molecular characterization of surface topology in protein tertiary structures by amino-acylation and mass spectrometric peptide mapping.* Bioconjug Chem. **1994**, 5, 583-90.

Florens L, Washburn MP, Raine JD, Anthony RM, Grainger M, Hayness JD, Moch JK, Muster N, Sacci JB, Tabb DL, Witney AA, Wolters D, Wu Y, Gardner MJ, Holder AA, Sinden RE, Yates JR, Carucci DJ. *A proteomic view of the Plasmodium falciparum life cycle.* Nature. **2002**, 419, 520-526.

Accurate m/z

592.4530 (obs.)
592.4513 (theo.) $C_{42}H_{58}NO$

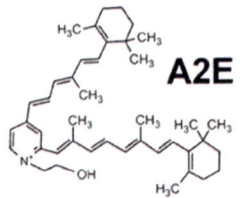

A2E

Retention Time

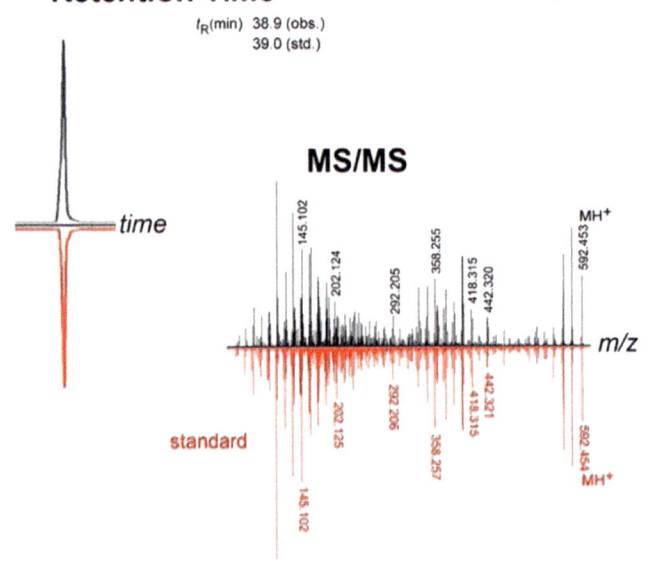

t_R(min) 38.9 (obs.)
39.0 (std.)

time

MS/MS

145.102
202.124
292.205
358.255
418.315
442.320
MH+ 592.453

standard

145.102
202.125
292.206
358.257
418.321
442.321
592.454 MH+

m/z

Chapter 6

Metabolomic Fundamentals

Perspective

The fundamental concepts of metabolomics and how mass spectrometry data acquisition is typically performed will be covered in this chapter. Historically speaking, I was personally intrigued to learn that the first GC/MS metabolite profiling was performed as early as 1966 by Dalgliesh et al., a good 27 years prior to my own efforts at Scripps in 1993 with LC/MS-based metabolite profiling on cerebral spinal fluid (before the term "metabolomics" was created). Times have changed, as metabolomics is now much more technologically and informatically sophisticated.

What is metabolomics and the metabolome?

The metabolome refers to the collection of small molecules produced by cells, which can provide valuable insight into the relationship between mechanistic biochemistry and cellular phenotype. Thanks to advances in mass spectrometry, it is now possible to measure thousands of metabolites simultaneously from only minimal amounts of sample, enabling comprehensive analysis without bias. Additionally, recent innovations in instrumentation, bioinformatic tools, and software have further facilitated these analyses and even provided for spatially localized detection of metabolites within biological specimens using mass spectrometry imaging. By applying these technologies, researchers have uncovered system-wide alterations of unexpected metabolic pathways in response to phenotypic perturbations, revealing a far more complex picture of cellular metabolism than was previously understood. Despite this progress, many of the molecules detected remain uncharacterized in databases and metabolite repositories, underscoring the incomplete nature of our current understanding of cellular metabolism. Nevertheless, the field of metabolomics has made tremendous strides in the last three decades since the first LC/MS profiling experiments were performed, and new tools are offering exciting mechanistic insights by enabling researchers to correlate biochemical changes with phenotype (**Figure 6.1**).

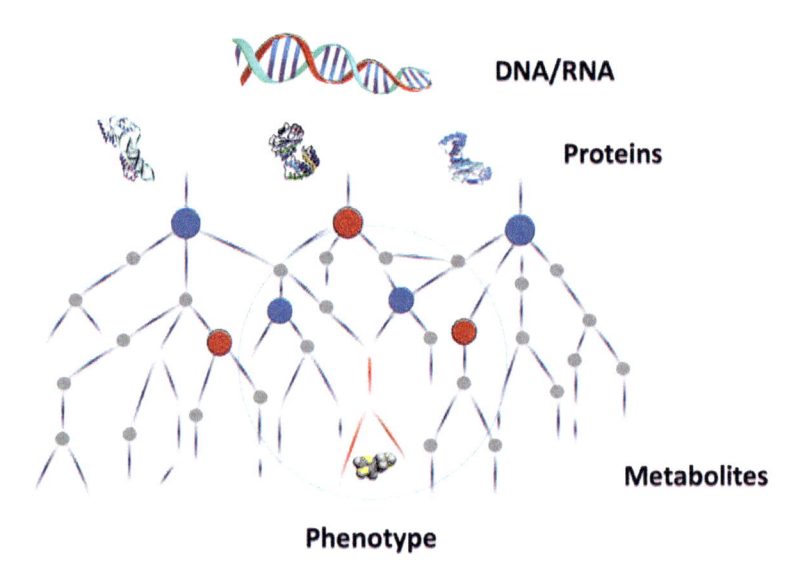

Figure 6.1. The central dogma of biology describes the flow of genetic information from DNA to RNA to protein. The 'omic cascade extends this flow to include the downstream products of cellular metabolism, namely metabolites and therefore closest to the phenotype. Alterations in a single gene or protein can lead to a cascade of metabolite alterations, illustrated by red and blue dots in the schematic. By performing meta-analysis (shown here with 3 different genetic alterations), metabolic alterations shared between multiple animal models or multiple genetic modifications may be identified, as shown by the superimposed Venn diagram. Up- and down-regulated metabolites are shown in red, and unaltered metabolites are shown in grey.

One topic of discussion will be the distinction between targeted and untargeted approaches to metabolomics. While both approaches have their merits, the value of untargeted metabolomics will be emphasized, and a practical guide to performing such an experiment will be provided. One of the first steps in any metabolomics experiment is to determine the number of metabolites to be measured. In some cases, researchers may choose to take a targeted approach and focus on a defined set of metabolites. In contrast, an untargeted or global approach may be preferred, in which as many metabolites as possible are measured and compared between samples without bias. The choice of approach will depend on the scientific question being addressed, and the number and chemical composition of metabolites studied will shape the experimental design, including decisions about sample preparation and choice of instrumentation. Ultimately, the approach chosen will have important implications for the depth and breadth of the metabolomic data generated and will influence subsequent data analysis and interpretation.

Targeted Metabolomics

Targeted metabolomics involves the measurement of a predefined set of metabolites, often focusing on one or more related pathways of interest. This approach is commonly driven by a specific biochemical question or hypothesis, motivating the investigation of a particular pathway. Targeted metabolomics is particularly effective for pharmacokinetic studies of drug metabolism and for measuring the influence of therapeutics or genetic modifications on a specific enzyme. Developments in mass spectrometry and nuclear magnetic resonance (NMR) have distinct advantages for performing targeted metabolomic studies, including their specificity and quantitative reproducibility. Although the term "metabolomics" is relatively recent, examples of targeted studies of metabolites date back to the 1960s, resulting in a wealth of literature investigating optimal protocols for the sample preparation and analysis of specific classes of metabolites. These protocols have been extensively discussed elsewhere.

Undoubtedly, targeted approaches have played a significant role in the development of metabolomics. Advances in using triple quadrupole (QqQ) MS (**Figure 6.2**) to perform selective reaction monitoring experiments have enabled routine analysis of most metabolites in central carbon metabolism, amino acids, and nucleotides at their physiological concentrations. These developments offer highly sensitive and robust methods to measure a significant number of biologically important metabolites with relatively high throughput. QqQ MS methods are quantitatively reliable, offering opportunities to achieve absolute quantitation of low-concentration metabolites that are difficult to detect with less sensitive methods such as NMR. Applying QqQ MS-based methods to human plasma allows screening of targeted lists of metabolites as potential metabolic signatures for disease.

Targeted Metabolomics with an LC triple quadrupole MRM

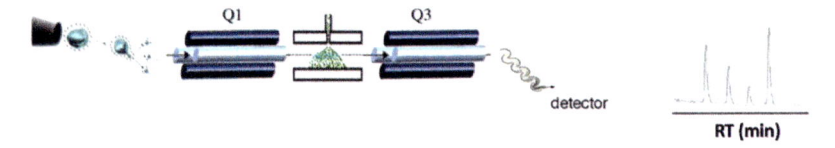

Untargeted (Global) Metabolomics with a QTOF or Orbitrap

Figure 6.2 Sample preparation and LC are similar with targeted and untargeted approaches, the primary distinguishing feature is the instrument used. For targeted, typically triple quadrupoles are used and for untargeted, either QTOFs or orbitraps.

Untargeted (or Global) Metabolomics

Untargeted metabolomic methods are designed to generate a global picture of the metabolome, providing for the simultaneous measurement of numerous metabolites from biological samples without bias. Although both NMR and MS technologies can be used to perform untargeted metabolomics, liquid chromatography followed by MS (LC/MS) is the preferred technique for global metabolite profiling. LC/MS-based metabolomic methods can routinely detect thousands of peaks, each of which represents a metabolite feature. These features correspond to a detected ion with a unique mass-to-charge ratio and retention time, although some metabolites may produce more than one feature.

Compared to targeted metabolomic results, untargeted metabolomic datasets are highly complex, with file sizes ranging in the order of gigabytes per sample for some of the latest high-resolution MS instruments. Manual inspection of the thousands of peaks detected is not practical and is further complicated by experimental drifts in instrumentation. For instance, deviations in retention time from sample to sample are common in LC/MS experiments due to factors such as column degradation, sample carryover, small fluctuations in room temperature, mobile phase pH, among others. Although these challenges initially presented significant obstacles for interpreting untargeted profiling data, remarkable progress has been made in the last decade through the development of metabolomic software such as XCMS and MZMine, making it routine to determine which features were dysregulated in global metabolomic datasets. These accomplishments have already provided insights into the fact that an astonishing number of metabolites remain uncharacterized regarding their structure and function, and that many of these uncharacterized metabolites change in response to health and disease. Hence, untargeted metabolomics has great potential to shed light on fundamental biological processes. In the rest of this article, we will focus on the untargeted metabolomic approach.

The impetus for untargeted metabolomics dates back to 1941, when G. Beadle and E. L. Tatum proposed the one gene–one enzyme hypothesis. This hypothesis was based on their experimental results, which showed that X-ray-induced mutant strains of the fungus Neurospora crassa were unable to carry out specific biochemical reactions. By systematically adding individual compounds to minimal N. crassa media and screening for those that rescued the growth of mutant strains, Beadle and Tatum identified metabolites whose biosynthesis had been affected by genetic mutation. Their work was the first to directly connect genotype to phenotype at the molecular level, and they purported that a single gene serves as the primary control of a single function, such as a specific chemical reaction. This hypothesis laid the foundation for the study of the relationships between genes and metabolic pathways and ultimately led to the development of modern untargeted metabolomics techniques.

Modern-day metabolomic experiments aim to connect genotype and phenotype by screening metabolites. However, the experimental screening methods used today are highly advanced, enabling the simultaneous study of many more compounds. Moreover, contemporary metabolic profiling experiments benefit from being complemented by genomic sequencing and proteomic screening. These global analyses have led to the emergence of systems biology, which has shown that a single non-lethal gene mutation can have dauntingly large effects. Indeed, single gene mutations can affect a significant number of metabolic pathways, thereby complicating the hypothesis that a single gene controls a single function. Furthermore, mutations in some unique genes can have unexpected phenotypic effects. For instance, the daf-2 gene encodes an insulin-like receptor in the nematode worm Caenorhabditis elegans. Mutations in daf-2 result in the worm living more than twice as long as its wild-type counterpart and cause alterations in the abundance of at least 86 identified proteins. Similarly, genes that encode for enzymes of the phosphatidyl-inositol 3-kinase family, which function in cell growth, proliferation, differentiation, motility, and signal transduction, are thought to have an oncogenic role in some cancers when mutated.

The examples provided illustrate how a single gene can have a far-reaching impact on numerous metabolic pathways and contribute to multiple cellular processes. Even with knowledge of protein structure, inferring function at the whole-organism level can be challenging due to complex regulatory mechanisms that involve epigenetic control, post-translational modifications, and feedback loops that enable context-dependent activation or deactivation. As a result, investigating the role of specific genes often requires systems-level analyses. Although these types of global studies were once limited to genes, transcripts, and proteins, recent technological developments have enabled untargeted profiling of metabolites and presented new opportunities to comprehensively track metabolic reactions directly.

Untargeted workflow. Untargeted metabolomic experiments are often hypothesis generating rather than hypothesis driven, so it is crucial to design experiments that maximize the detection of metabolites and ensure their quantitative reproducibility. However, metabolite identification can be a time-consuming process, so it is important to carefully choose the sample type, preparation, chromatographic separation, and analytical instrumentation to obtain high-quality data. In this context, LC/MS-based workflows are preferred because they offer high sensitivity and enable the detection of a large number of metabolites using minimal sample amounts (typically less than 25 mg of tissue, approximately 1 million cells, or around 50 µL of biofluids such as plasma and urine). Therefore, it is important to consider and choose the workflow that is most likely to yield high-quality data for analysis.

Sample preparation and data acquisition. The first step in untargeted metabolomic workflow is to extract metabolites from biological samples. Several approaches, such as sample homogenization and protein precipitation, have been utilized and are described in detail elsewhere. Prior to MS analysis, the isolated metabolites are chromatographically separated using relatively short solvent gradients, enabling high-throughput analysis of large sample numbers. Multiplexing, extraction, and separation methods maximize the number of metabolites detected due to the heterogeneous physiochemical landscape of the metabolome. For instance, extracting the same cells with both organic and aqueous solvents increase the detection of hydrophobic and hydrophilic compounds, respectively. Similarly, reversed-phase chromatography is better suited for separating hydrophobic metabolites, while hydrophilic-interaction chromatography is generally more effective for separating hydrophilic compounds. Although quadrupole time-of-flight (QTOF) mass spectrometers and Orbitrap mass spectrometers are most commonly used to collect data, other instruments such as time-of-flight and ion trap instruments can also be used. Because predicting tandem MS (MS/MS) fragmentation patterns for most metabolites is challenging, unlike in shotgun 'omic approaches, untargeted metabolomic profiling data is typically acquired in MS1 mode, measuring only the mass-to-charge ratio (m/z) of the intact metabolite.

Data analysis. Advances in bioinformatic tools have led to a more automated process for identifying metabolite features that are differentially altered between sample groups. Various metabolomic software programs are available that provide a range of capabilities, including peak picking, non-linear retention time alignment, visualization, relative quantitation, and statistical analysis. Among the available software programs, XCMS is a widely used and is freely accessible online. Users can upload data and perform data processing, as well as browse results within a web-based interface. (https://xcmsonline.scripps.edu/).

Metabolite identification. It is worth noting that current metabolomic software does not effectively output all metabolite identifications. Instead, it provides a table of features with p-values and fold changes that indicate their relative intensity differences between samples. To determine the identity of a feature of interest, the accurate mass of the compound is first searched in metabolite databases like METLIN. However, a database match represents only a putative metabolite assignment that must be confirmed by comparing the retention time and MS/MS data of a molecular standard to that of the feature of interest in the research sample.

Metabolite Characterization

Accurate m/z
592.4530 (obs.)
592.4513 (theo.) $C_{42}H_{58}NO$

A2E

Retention Time
t_R(min) 38.9 (obs.)
39.0 (std.)

time

MS/MS
145.102
202.124
292.205
358.255
418.315
442.320
MH⁺ 592.453

m/z

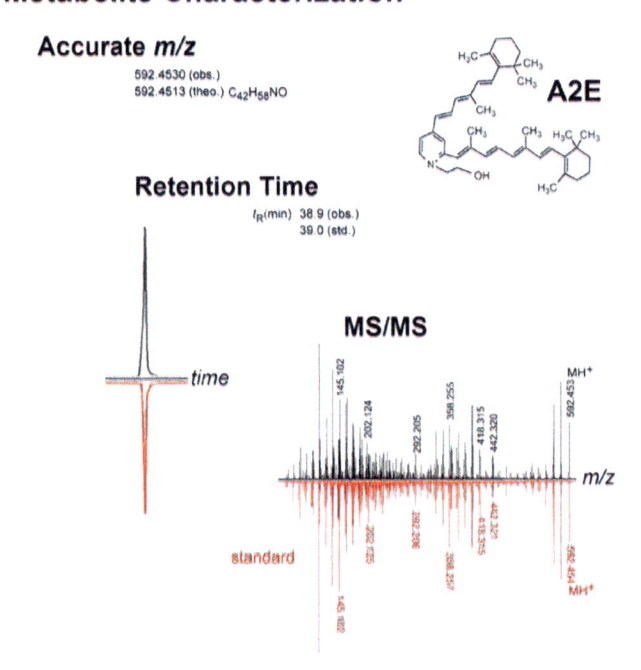

standard
145.038
292.225
362.225
398.357
418.315
456.454
MH⁺

Figure 6.3 Metabolite identification of A2E (A2-ethanolamine). Accurate m/z of the compound is measured and compared to the theoretical m/z with an error of less than 3 ppm. Second, the retention time (38.9 min, black) is compared to that of a standard (39.0 min, red). Finally, the MS/MS spectra (black) and the standard (red) match.

Currently, the additional experiments required to obtain MS/MS data for selected features from the profiling results, as well as the matching of MS/MS fragmentation patterns, are performed manually by inspection (**Figure 6.3**). These analyses are time-intensive and represent the rate-limiting step of the untargeted metabolomic workflow. Moreover, despite the growth of metabolite databases over the last decade, a significant number of metabolite features detected from biological samples do not return any matches. Identifying these unknown features requires de novo characterization with traditional methods. It is important to recognize that comprehensive identification of all metabolite features detected by LC/MS is currently impractical for most samples analyzed.

Challenges

The field of metabolomics has seen remarkable advancements in recent years, thanks to the development of innovative technologies and experimental strategies that have enabled researchers to better understand the complex and dynamic nature of the metabolome. Untargeted metabolomics, in particular, has revolutionized the field by revealing that the number of endogenous metabolites in biological systems is much larger than previously thought.

Despite these advances, however, there are still many challenges that need to be addressed. One of the main challenges is the fact that the metabolome is not encoded in the genome, which makes it difficult to analyze an undefined set of molecules. To overcome this challenge, metabolite databases have been rapidly expanding, which has facilitated untargeted studies.

However, despite the expansion of metabolite databases, there are still many metabolites for which the chemical structure, cellular function, biochemical pathway, and anatomical location remain uncharacterized. To address this issue, researchers are increasingly turning to innovative technologies and experimental strategies that can be coupled with untargeted profiling.

For example, the development of high-resolution mass spectrometry (HRMS) has enabled researchers to detect a wider range of metabolites, including those with unusual or unexpected masses. Additionally, the use of stable isotope labeling has allowed researchers to trace the fate of specific metabolites in biological systems, providing valuable insights into their metabolic pathways and cellular functions.

Other innovative strategies that are driving progress in the field include the use of multi-omics approaches, which integrate metabolomics data with other omics data, such as genomics, transcriptomics, and proteomics, to provide a more comprehensive understanding of biological systems. Furthermore, the use of machine learning algorithms and artificial intelligence (AI) is enabling researchers to analyze and interpret large amounts of metabolomics data, facilitating the identification of novel biomarkers and metabolic pathways.

Improving metabolite databases. Over the last two decades, the information catalogued in metabolite databases has evolved beyond lists of one-dimensional data that is traditionally acquired by tandem mass spectrometry. The NIST tandem mass spectrometry database, for example, now stands at 31K molecules with MS/MS data. However the METLIN database has leapfrogged all other databases and contains tandem mass spectrometry experimental data for over 900,000 molecular standards. All METLIN data has been generated from molecular standards and has been acquired in both positive and negative ionization modes and at four different collision energy (in each ionization mode).

Meta-analysis: Altering a single enzyme can set off a chain reaction of metabolic perturbations that may not be related to the intended phenotype. Therefore, when conducting untargeted metabolomic profiling of a particular disease or mutant, hundreds of alterations may be observed that lack mechanistic significance. Given the considerable resources required to identify both known and unknown compounds, it is advantageous to use strategies that can reduce the list of potentially interesting features before committing time to identification. One such strategy is meta-analysis, whereby untargeted profiling data from multiple studies are compared (**Figure 6.1**). For example, by comparing different models of a disease, features that are not similarly altered in each comparison can be given lower priority as they may be less likely to be related to the shared phenotypic pathology. A freely available software called metaXCMS has been developed to automate the comparison of untargeted metabolomic data. As a proof of concept (**Figure 6.4**), metaXCMS was applied to investigate three pain models of varying pathogenic etiologies: inflammation, heat, and arthritis.

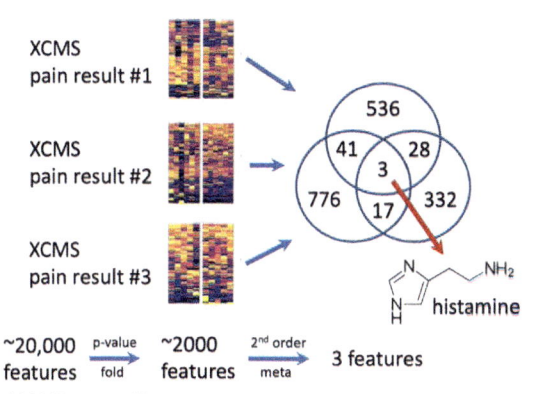

Figure 6.4, metaXCMS was utilized to investigate three different pain models, which have different sources: inflammation, heat, and arthritis. The analysis revealed hundreds of metabolite features that were dysregulated in each model. However, only three metabolites were found to be commonly dysregulated among all the groups. One of the shared metabolites was histamine, a well-known mediator of pain that functions through multiple mechanisms. This finding highlights the potential of 'similar data-reduction' strategies to identify physiologically relevant unknown features in other biological systems.

While hundreds of metabolite features were found to be altered in each model, only three were similarly dysregulated among all groups. One of the shared metabolites was histamine, a well-characterized mediator of pain that functions through several mechanisms. By applying similar data-reduction strategies to other biological systems, it may be possible to justify aggressive analytical investigations of physiologically relevant unknown features.

Imaging approaches to localize metabolites. Metabolite isolation through sample homogenization is a crucial initial step in the untargeted metabolomic workflow for analyzing biological tissues. However, the lack of high-resolution spatial localization of metabolites within samples, due to standard metabolic profiling techniques, poses a significant challenge. The averaging of different cell types with potentially unique metabolomes in heterogeneous tissues, such as the brain, further complicates investigations. As a result, identifying a dysregulated metabolite and its specific location within a tissue or cell type can be challenging. To overcome these limitations, advanced imaging techniques and analytical methods, such as mass spectrometry imaging, are being developed to enable more precise spatial resolution of metabolites in tissues.

Mass Spectrometry Imaging (MSI) of [Cholesterol + Ag]⁺

493.3

m/z

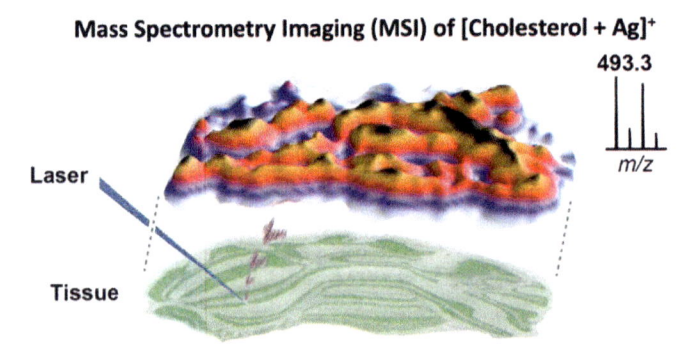

Laser

Tissue

Figure 6.5. Spatial localization of metabolites in tissue by mass spectrometry-based imaging. Example of a surface-based image of [cholesterol + Ag]+ from mouse brain by using nanostructure-imaging mass spectrometry (NIMS). NIMS is well-suited for metabolite imaging because it is sensitive and does not suffer from matrix interference in the low-mass range. Sections of frozen tissue are first transferred to a NIMS chip that is subsequently analyzed by using a laser-induced desorption/ionization approach. By systematically rastering the laser across the tissue, a mass spectrum is generated from each point.

MS-based imaging relies on matrix-assisted laser desorption ionization (MALDI) offer chemical specificity and sensitivity, but they are limited in their application to metabolites due to background interference caused by the matrix in the low-mass region characteristic of metabolites. As an alternative, a matrix-free technique called nanostructure imaging mass spectrometry (NIMS) has been developed for the analysis of metabolites with high sensitivity and spatial resolution (**Figure 6.5**). By using NIMS to analyze 3 μm sections of brain tissue from mice with impaired cholesterol biosynthesis, metabolic precursors of cholesterol were found to localize to the cerebellum and brainstem. These types of NIMS imaging applications coupled with histology will allow metabolite localization patterns to be correlated with tissue pathology and drive developments in our understanding of chemical physiology.

Overview

Over the years, there has been a growing interest in metabolic profiling, yet only in recent times have technologies emerged that enable the global analysis of metabolites at a systems level comparable to genomics, transcriptomics, and proteomics. Unlike its 'omic counterparts, metabolomics allows for the direct measurement of biochemical activity by monitoring the substrates and products involved in cellular metabolism. This untargeted profiling of chemical transformations at a global level provides a phenotypic readout that has been useful in clinical diagnostics, identifying therapeutic targets of diseases, and investigating the mechanisms of fundamental biological processes.

Although untargeted metabolomics is still in its early stages, recent studies have shown that the complexity of comprehensive cellular metabolism is more extensive than what classical biochemical pathways suggest. This realization is reminiscent of how the emergence of experimental results, such as the photoelectric effect, in the early twentieth century challenged Newtonian laws and eventually led to the development of quantum mechanics. As metabolomic technologies continue to advance and enable the characterization of unknown pathways, untargeted metabolomics has the potential to shape our understanding of global metabolism.

References

Johnson, C., Ivanisevic, J. & Siuzdak, G. Metabolomics: beyond biomarkers and towards mechanisms. *Nat Rev Mol Cell Biol* **17**, 451–459 (2016). https://doi.org/10.1038/nrm.2016.25

Patti, G., Yanes, O. & Siuzdak, G. Metabolomics: the apogee of the omics trilogy. *Nat Rev Mol Cell Biol* **13**, 263–269 (2012). https://doi.org/10.1038/nrm3314

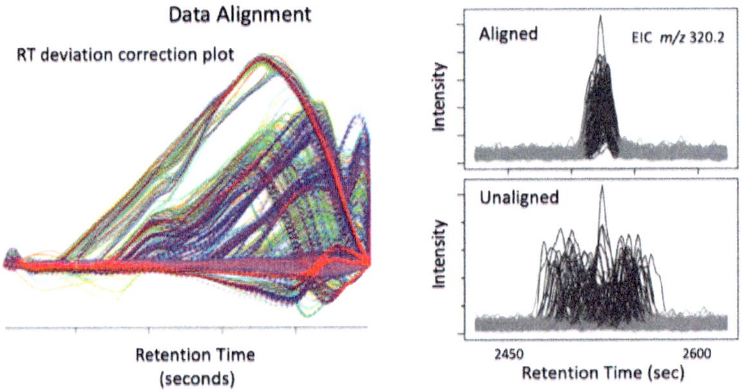
Data Alignment
RT deviation correction plot
Retention Time
(seconds)
Aligned
EIC m/z 320.2
Intensity
Unaligned
Intensity
2450
2600
Retention Time (sec)

Chapter 7
XCMS Data Processing

Perspective

XCMS (eXtensible Computational Mass Spectrometry) is a well-known software platform utilized for the processing of liquid chromatography–mass spectrometry (LC-MS) data, available both as an R package and a cloud-based platform called XCMS Online. It is especially known for introducing nonlinear alignment for LC-MS data processing (**Figures 7.1 and 7.2**), thus facilitating more accurate statistical comparisons across data sets. This chapter provides an overview of LC-MS data analysis, emphasizing the importance of data processing software. Subsequently, we delve into the algorithms employed by XCMS and elucidate the significance of user-defined parameters in LC-MS data processing. Lastly, we explore the additional functionalities provided by XCMS Online, thereby showcasing its expanded capabilities.

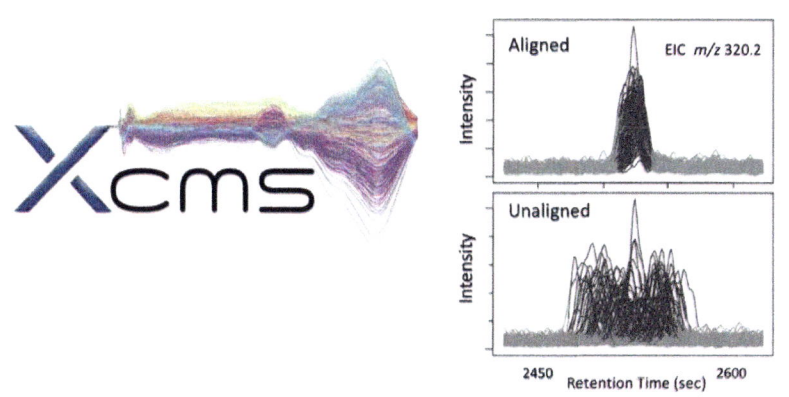

Figure 7.1. The logo that has come to represent XCMS (right) was derived from the very first nonlinear alignment of LC/MS data (right) using an algorithm created by Colin Smith. The beauty of that original compilation of LC/MS correction plots led me to use the plot routinely, which has since become synomonous with XCMS.

How are metabolomics data processing platforms used?

Untargeted or global metabolomics is a research approach focused on quantifying a wide range of metabolites in biological samples. Among various technologies employed for this purpose, liquid chromatography–electrospray ionization mass spectrometry (LC-ESI-MS or simply LC-MS) stands out as one of the most commonly used methods. This popularity is owed to its straightforward sample preparation process and extensive coverage of metabolites. LC-MS generates vast and intricate datasets, consisting of thousands of metabolic features. Consequently, computational tools are crucial for effectively processing this data and transforming it into meaningful and interpretable information.

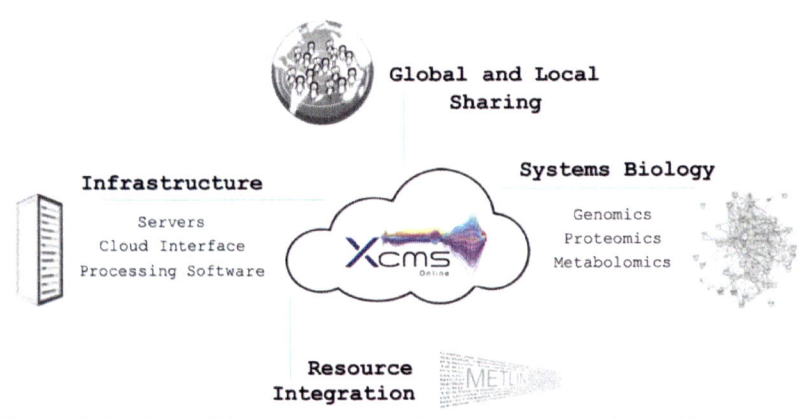

Figure 7.2. Cloud-based XCMS computing provides an interconnectedness with infrastructure, data sharing and the ability to integrate multiple resources. The multi-omics XCMS Online platform with over 50,000 registered users represents a freely available, cloud-based resource offering data processing, archiving and sharing, easy-to-use statistical tools, intuitive visualization, pathway analysis and metabolite identification.

The untargeted computational data processing workflow (**Figure 7.3**) gained significant traction in 2006 following the publication of XCMS. This workflow typically involves a series of steps, starting with peak picking and nonlinear alignment. The outcome of this process is a collection of features, representing individual peaks or groups of peaks with distinct *m/z* values and retention times across samples. The objective is to convert the raw data into a matrix that comprises

7. XCMS Data Processing

a list of observed features, accompanied by their corresponding relative peak areas or intensities for each sample in a given experiment. This compilation of features, known as the feature list, enables the comparison of peak areas (correlating to relative concentrations) among identical ion peaks across samples, facilitating the identification of statistically significant dysregulated peaks associated with a particular phenotype.

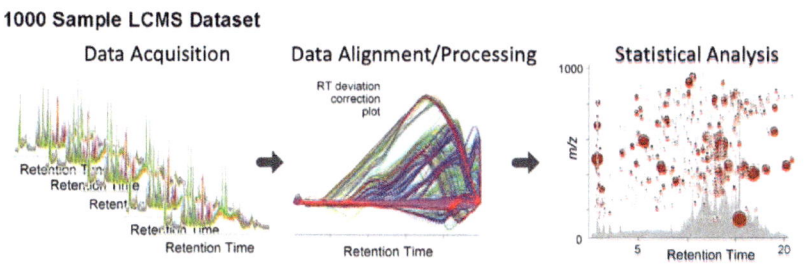

Figure 7.3. XCMS workflow from LC/MS data acquisition to data nonlinear alignment to statistical analysis (here represented in a cloud plot). In the cloud plot the size of the bubble reflects fold change and the intensity of the color reflects p-value. Information for each metabolic feature is also tabulated, including direction of dysregulation, fold change, P value, m/z, retention time, adduct form and the XCMS feature ID number. Clicking the unique feature ID (or bubble on the cloud plot) opens a pop-up window displaying the MS spectrum, LC chromatogram and box-and-whisker plot for that metabolic feature.

XCMS Data Streaming

A unique feature of XCMS is its implementation of data streaming, addressing the issue that data transfer times can represent a bottleneck in data processing due to the increasingly complex data files including greater number of samples. To meet the demand of analyzing hundreds to thousands of samples, we developed data streaming which capitalizes on the acquisition time to stream recently acquired data files to data processing servers, mimicking just-in-time strategies from the entertainment industry. Its utility has been demonstrated with as much as a 10,000-fold time savings, dramatically reducing data analysis time (**Figures 7.4**).

Streaming allows users to increase the efficiency of their analytical workflow by reducing data transfer times. Previously, data acquisition time was considered "useless" in terms of data processing and analysis time. However, streaming data files during acquisition can provide significant time savings (hours to days).

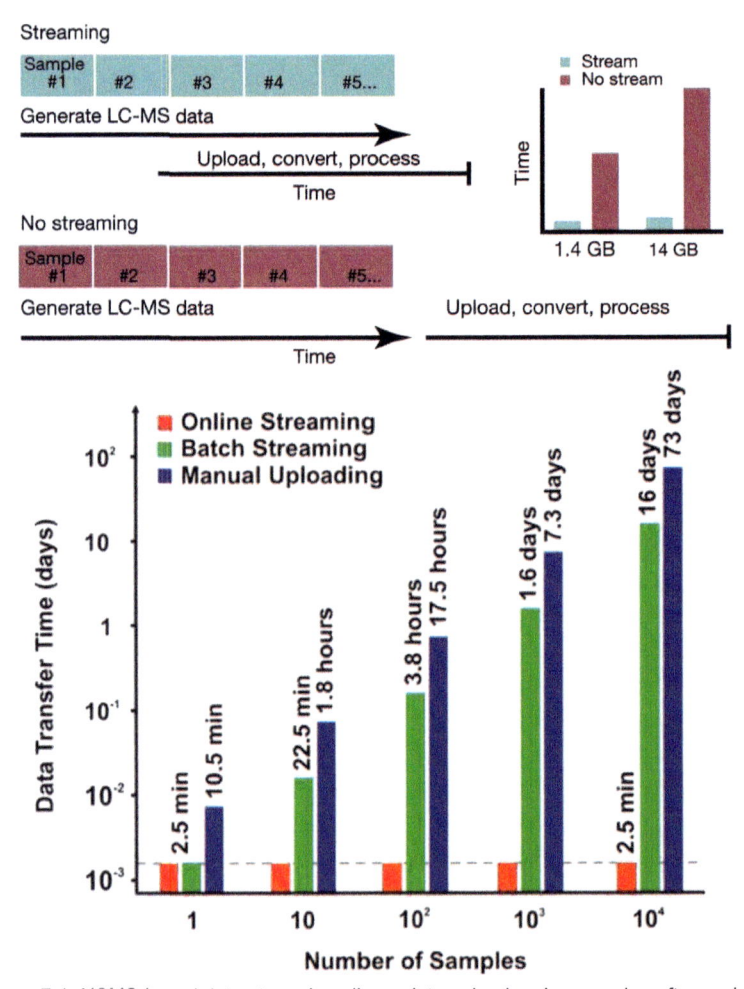

Figure 7.4. XCMS-based data streaming allows data upload and processing after each LC-MS run is performed, reducing the processing time after the data are acquired for the final sample. Time comparison between streaming and manual data uploading. Data transfer time comparison in days (logarithmic scale) for different number of samples between online, batch streaming and manual uploading.

7. XCMS Data Processing

XCMS Data Analysis Example

Following the steps of data acquisition, peak-picking and nonlinear alignment, subsequent processes in untargeted metabolomics typically involve statistical analysis to identify dysregulated peaks associated with a specific phenotype, as well as metabolite annotation/identification. **Figure 7.5** provides an overview example of an experiment where untargeted metabolomic data was used to understand how arterial and venous blood differs.

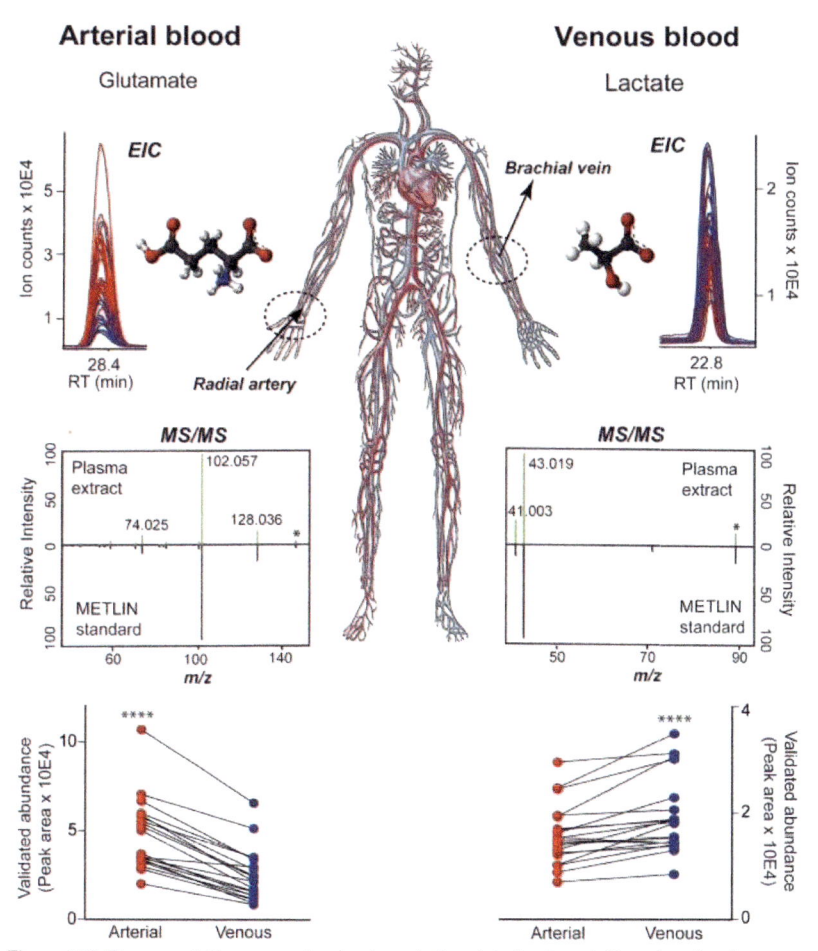

Figure 7.5. Representative example of untargeted metabolomics, Julijana Ivanisevic analyzed metabolites in human arterial and venous blood, both collected from the same subjects. EICs show glutamate higher in arterial blood while lactate was higher in venous (unsurprisingly as it is a muscle waste product).

181

Computational metabolite annotation aims to provide chemical information about the observed features. This annotation process typically entails determining which features correspond to the same metabolite, characterizing the nature of the features (e.g., discerning if a feature represents a protonated/deprotonated species, a cation or anion adduct, an in-source fragment, an isotope, a dimer, etc.), and offering putative metabolite identifications.

An example of the basic annotation process involves comparing the m/z values of the observed features with a database to determine their potential identity. However, this approach is considered relatively inaccurate, and in the forthcoming chapter, we will delve into more advanced techniques that utilize fragmentation information however **Figure 7.5** provides a representative example of how intact m/z data of the precursor ion combined with fragmentation data from MS/MS experiments allow for identification of glutamate and lactate.

XCMS Cloud Plot

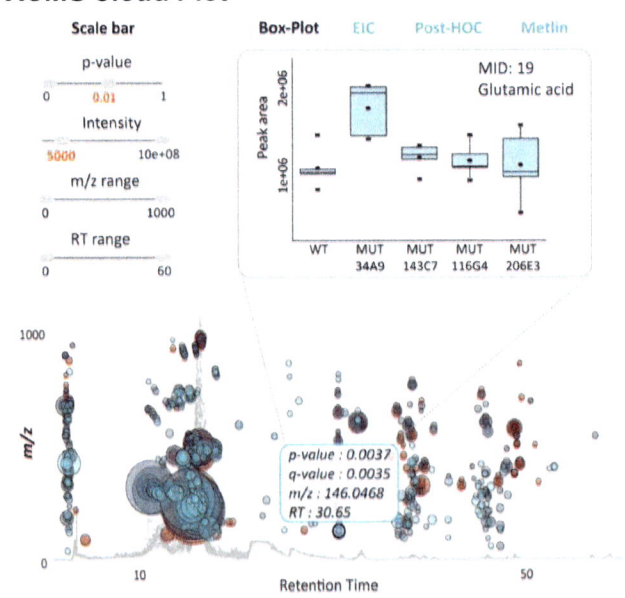

Figure 7.6. A cloud plot with each circle representing a feature that was dysregulated, the size and the intensity of the color represent fold change and p-value, respectively. Additional data can be accessed by clicking on each circle.

7. XCMS Data Processing

XCMS also has a unique display feature in it cloud plot, the cloud plot (**Figures 7.6 and 7.7**) provides bubbles or circles that represent dysregulated features within a data set. The size of the bubble denotes fold change and the intensity of the color denotes p-value. How the values are displayed can be adjusted within the scale bar (**Figure 7.6**) to filter based on p-value, intensity, m/z range and RT range, at your discretion.

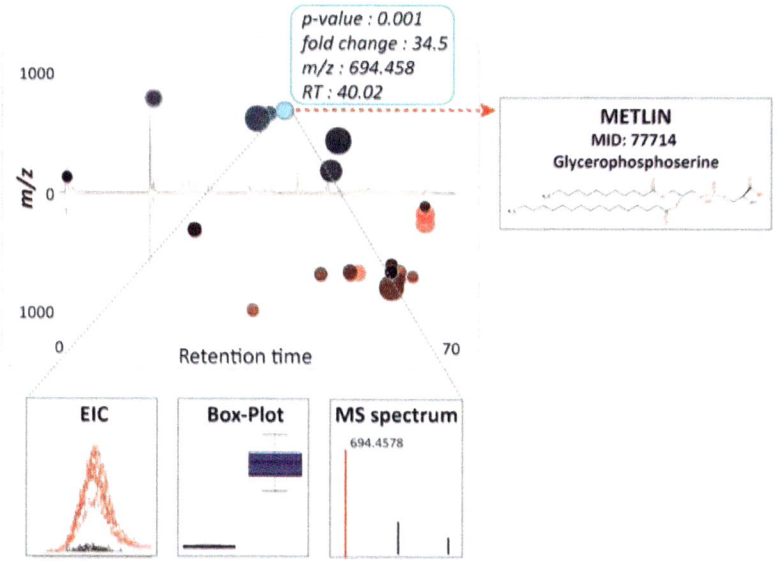

Figure 7.7. Interactive cloud plot with customized metabolomic data visualization. When a user scrolls the mouse over a bubble, feature assignments are displayed in a pop-up window (m/z, RT, p-value, fold change) with potential METLIN hits. When a bubble is selected by a mouse click, its EIC, box–whisker plot, and MS spectrum appear on the bottom of the main panel.

XCMS Meta Analysis

Another feature of XCMS is meta analysis (**Figures 7.6 and 7.8**). Mass spectrometry-based untargeted metabolomics often results in the observation of hundreds to thousands of features that are differentially regulated between sample classes. A major challenge in interpreting the data is distinguishing metabolites that are causally associated with the phenotype of interest from those that are unrelated but altered in downstream pathways as an effect. To facilitate this distinction, meta-analysis has been incorporated within XCMS for performing second-order ("meta") analysis of untargeted metabolomics data from multiple sample groups representing different models of the same phenotype.

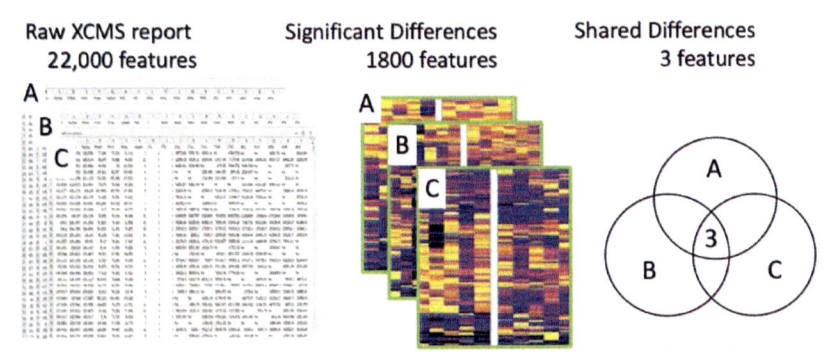

Figure 7.8. Data reduction by meta-analysis. (Top) Three pairwise comparisons of different pain models with their respective controls resulted in 22,577 detected metabolite features, significant differences 1,825 features and shared differences 3 features, one of which was determined to be histamine. The putative identity of those metabolites, is verified by MS/MS matching to standards in METLIN.

While the original version of XCMS was designed for the direct comparison of two sample groups, meta-analysis enables an unlimited number of sample classes to facilitate prioritization of the data and increase the probability of identifying metabolites causally related to the phenotype of interest. Meta-analysis across multiple datasets is filtered, aligned, and ultimately compared to identify shared metabolites that are up- or down-regulated across all sample groups.

What's Going on Inside XCMS

The XCMS computational workflow combines peak picking with alignment and is accessible as both an R package and a cloud-based resource. The cloud-based version provides an intuitive visual interface, enabling users to share data with the scientific community or collaborators. Moreover, it encompasses a comprehensive suite of tools that facilitate systems biology analysis and metabolite annotation. Notably, XCMS now integrates with the METLIN spectral library (Chapter 9), enabling improved annotation capabilities. Additionally, a targeted metabolomics version of XCMS Online, known as XCMS-MRM, is also available for focused metabolomics investigations.

In LC-ESI-MS, each identified molecule generates multiple chromatographic peaks. Typically, when employing common untargeted metabolomics chromatographic methods, a single molecule will generate a protonated or deprotonated species. Additionally, various phenomena like isotopes, adducts, and in-source fragments may also be observed. Notably, in-source fragmentation, a frequent occurrence in LC-ESI-MS, leads to the detection of fragments or neutral losses. Some of these neutral losses, known as common neutral losses (e.g., -H_2O, -NH_3, -HCOOH), are observed across a broad range of organic molecules found in biological systems. Less commonly observed peaks include dimers or multimers. Overall, LC-MS generates complex datasets typically report tens of thousands of metabolite features detected. The effective attribution of all these peaks to bona fide metabolites remains an active area of research.

XCMS Workflow

XCMS serves as a computational tool that combines peak picking and alignment to generate a comprehensive list of features from raw MS data (**Figures 7.9**). Peak picking is a computational procedure employed to detect peaks in the MS data and calculate their integrated area. Traditionally, two types of peak picking algorithms have gained popularity: the match filter algorithm and its advanced variation for high-resolution mass spectrometry data, known as the centWave algorithm. XCMS provides the option to utilize either the original match filter algorithm or the centWave algorithm. In this section, we will delve into the details of these two algorithm types.

Peak Detection → Peak Grouping → Fill Peaks → Statistical Analysis

↓↑

Peak Alignment

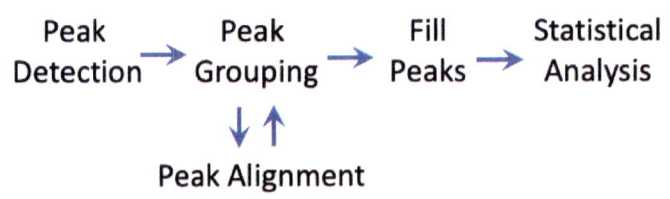

Figure 7.9. XCMS general workflow. The workflow consists of peak detection, grouping of peaks into features, alignment, missing peaks filling, and the statistical analysis of features across classes.

When peaks are detected in raw MS data across multiple samples, it is desirable for the same peak originating from a specific ion to appear consistently across the different samples, albeit with varying intensities. Modern mass spectrometers, such as TOF and Orbitrap instruments, offer high accuracy, resulting in m/z values exhibiting variations as low as 1 ppm across samples. Conversely, the retention time tends to have larger variations among samples. Peaks from the same ion detected in different samples need to be aligned and grouped, which involves assigning their respective areas to the same row in a data matrix, enabling quantitative comparisons across samples. XCMS accomplishes this peak alignment through a two-step process referred to as peak grouping and retention time alignment, which we will elucidate in the following sections.

Once peaks are detected and aligned, statistical analyses can be employed to identify features that display statistically significant changes in peak areas between different phenotypes. However, following peak picking and alignment, it is common to encounter missing values, which affect more than 80% of the detected features. This implies that approximately 80% of the detected features will have missing values for certain samples, indicating zero peak area or intensity. Such occurrences arise when the ion peak falls below the detection limit in a given group of samples. In other cases, the peak may be observed above the detection limit, but the peak picking algorithm fails to detect it due to low intensity or interference from noise or coeluting peaks. The presence of missing values reduces the power of statistical tests and analyses and can lead to biased results. This issue becomes especially pertinent when conducting multivariate analyses, as missing values can introduce bias into the results. To address this problem, various strategies, collectively known as missing value imputation strategies, have been devised. One widely accepted approach is known as "filling peaks," wherein "missing" peaks are re-examined in the raw data. XCMS adopts a fill peaks strategy to "fill" the "missing" peaks in the feature list, and we will provide a detailed explanation of this strategy in the subsequent sections.

The subsequent sections explain the different steps comprising the XCMS workflow, which include (1) peak picking, (2) peak grouping and retention time alignment, and (3) fill peaks.

XCMS Workflow - Peak Picking

Peak picking is a crucial step in identifying the chromatographic peaks originating from molecules eluting from the chromatographic column. Traditionally, peak picking algorithms have been based on filtering techniques that remove noise from the signal, enabling the identification of distinct peaks within the raw data. Numerous peak picking algorithms have been developed, some tailored for specific types of data. In the case of LC-MS data, peak detectors must process the complex three-dimensional LC-MS data to detect these peaks. Certain algorithms, like the "centWave" algorithm, require data reduction prior to analysis. This reduction involves converting the data, typically acquired in profile mode, into centroid mode data. Conversion can significantly reduce the data size and simplify subsequent processing. While most instrument vendors provide their own centroiding algorithms, open-source alternatives are also available. However, caution should be exercised when using these alternative algorithms to retain mass accuracy.

In XCMS, two algorithms, namely Matched Filter and centWave, can be employed for peak picking.

XCMS Workflow - Matched Filter Peak Picking

The original peak detection algorithm of XCMS follows a step-by-step process. First, the algorithm employs a "binning" procedure to divide the data into bins of 0.1 m/z units (determined by the step parameter). Within each bin, peaks are identified by surpassing the signal-to-noise threshold specified by the S/N ratio cutoff parameter (default value: 10). The algorithm takes advantage of the characteristic Gaussian-like shape of chromatographic peaks to identify them. Data points that conform well to a second derivative Gaussian (resembling a normal distribution) are classified as peaks. The width of this Gaussian is determined by the full width at half maximum (FWHM) parameter, expressed in seconds. Lower FWHM values increase the likelihood of false positive peak detection, where noise is erroneously identified as a peak. To prevent peaks from being split between adjacent m/z bins, the algorithm combines pairs of consecutive bins. Originally designed for data acquired by low-resolution instruments like single quadrupole mass spectrometers with a maximum mass accuracy of approximately 0.1 Da, the centWave algorithm is recommended for current high-resolution mass spectrometers (HRMS) (explained below). However, if the MatchedFilter algorithm is applied to HRMS data, the binning value should be significantly reduced. The output of the algorithm is a peak list containing the m/z and retention time (along with their deviations) as well as the integrated peak intensity (peak area) for each identified peak.

XCMS Workflow - CentWave Peak Picking

First introduced in 2008, the "centWave" algorithm is specifically designed for high-resolution mass spectrometry, offering superior mass accuracy peak detection. This algorithm comprises two steps to identify peaks. In the initial step, a dynamic binning approach is applied to identify regions of interest (ROIs) that potentially contain peaks (**Figure 7.10**). The subsequent step involves peak detection within these ROIs using a highly sensitive wavelet filter. The algorithm searches for ROIs characterized by low m/z deviations and intensity profiles that exhibit an increase and subsequent decrease. Key parameters influencing this behavior include the "ppm" parameter, which defines the m/z range of the ROIs, and the peak width, measured in seconds, which determines the duration of the peak in the chromatographic time domain. Once a qualifying ROI is identified, it undergoes analysis by the wavelet filter. The Mexican

7. XCMS Data Processing

hat wavelet is employed as the model for the peak shape, enabling the detection of multiple closely eluting peaks within the ROI. The peak scale or height is adjusted until the best fit is achieved. If the fitting criteria are not met, the ROI is rejected as a peak. The output of the centWave algorithm is comparable to that of the MatchedFilter, providing information on the m/z, retention time (including deviations), and integrated intensities of the detected peaks.

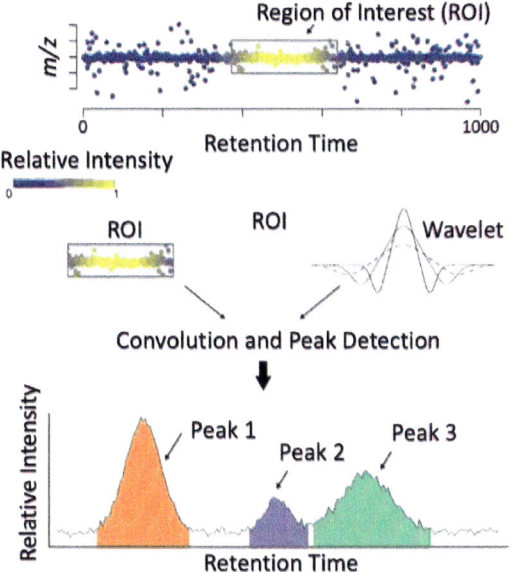

Figure 7.10. CentWave peak detection overview. First, regions of interest (ROI) are detected. For each detected ROI, a wavelet filter is applied and peaks within the ROI are detected.

XCMS Workflow - Peak Alignment

Alignment In XCMS, peak alignment (**Figure 7.11**) involves a two-step process: peak grouping and alignment across samples. The order in which these steps are executed depends on the alignment algorithm used.

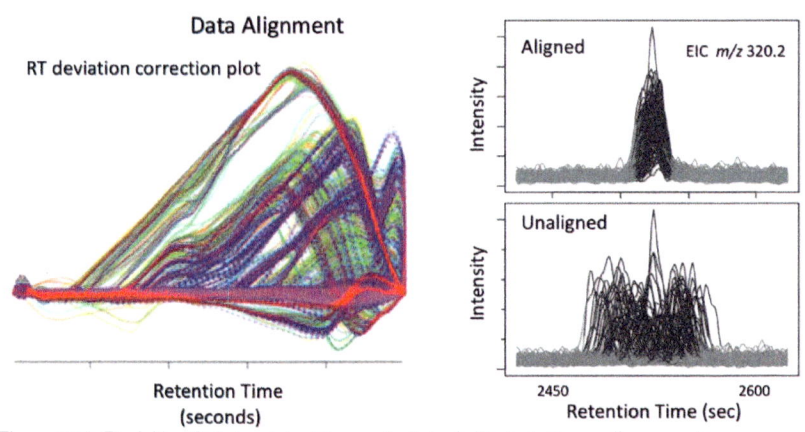

Figure 7.11. Revisiting Colin's original figures that nicely illustrate the nonlinear peak alignment impact on the data, and therefore improving statistical analysis.

When the original "peakGroups" alignment method is selected, peaks are initially grouped into features. This grouping is performed by the "peakDensity" algorithm, which operates on the processed peak list rather than the raw data. Utilizing thin m/z slices (defined by the "mzwid" parameter), the algorithm identifies groups of peaks across samples that cluster around a specific retention time (RT). The objective is to identify "well-behaved" peak groups (WBPG) that can serve as a basis for aligning the remaining chromatograms, thereby enhancing the accuracy of peak grouping. To identify these WBPG, the algorithm employs a kernel density filter that identifies regions with clusters of "well-behaved" peaks. The filter's behavior can be adjusted using the "bw" parameter or bandwidth, which is defined in seconds. For HPLC data, a common default value is 30 seconds. WBPG should be distributed throughout the retention time to facilitate effective nonlinear correction. Using the median retention time of each WBPG, an alignment profile is created using a regression technique called LOcally Estimated Scatter-plot Smoother (LOESS). LOESS fits a smoothed line through each chromatogram and compares these smoothed lines across chromatograms to correct the retention time across samples.

An alternative to the nonlinear retention time alignment algorithm is the "obiwarp" algorithm, introduced. Unlike the pregrouping approach, "obiwarp" directly aligns peaks across samples. This algorithm employs dynamic programming to determine the best alignment between two chromatograms. It operates in a pairwise manner, first establishing a median retention profile across multiple chromatograms and then iteratively aligning all chromatograms to that median profile. It is worth noting that the "obiwarp" algorithm can be computationally intensive and time-consuming.

Regardless of the chosen retention time alignment algorithm, the subsequent step involves grouping the peaks to generate features, which are defined as a single or a set of peaks across samples with a unique RT and m/z. The aforementioned "peakDensity" algorithm can be employed once again to accomplish this task. With the peaks now aligned, the "bw" parameter can be significantly reduced. As an additional robustness filter, the "minfrac" parameter comes into play. This parameter specifies that, for any given class (e.g., KO for knockout, WT for wild type), there must be a minimum fraction of peaks present in a feature for it to be considered valid. For example, if there are 6 samples in the KO class and 6 samples in the WT class, and the "minfrac" is set at 50%, there must be at least 3 peaks present in either class for a particular m/z and RT combination to be considered a valid feature. It's important to note that it applies to "any one class" and not both classes. Therefore, even if KO has 3 peaks and WT has 0, it would still be deemed a valid feature. This parameter can enhance robustness when searching for peaks across all samples.

XCMS Workflow - Fill Peak

As mentioned earlier, there are two primary reasons why peaks may be missed. Firstly, the algorithm might fail to correctly detect or align the peak, or the peak could be below the detection limit. The fill peaks step effectively addresses these issues by utilizing information from other samples where the peak has been successfully detected (retention time and m/z). When the problem lies with the algorithm, the fill peaks step will accurately identify and recover missing peaks by leveraging the data from these reference samples. In cases where the peak is below the detection limit, the fill peaks step employs the background noise in the corresponding region to estimate the missing peak value. It is worth noting that a peak's absence in one class of samples and its presence in others may have biological implications. Therefore, the user has the flexibility to define the minimum number of peaks required to be found within a single class, accommodating such scenarios.

XCMS R Package

XCMS, an open-source R package, has an active community that continuously updates and enhances its capabilities. The R version of XCMS provides a command line interface, allowing users to directly interact with their data using a set of commands. In this chapter, we have included workflows for both the original and alternative pipelines.

The latest version of XCMS, referred to as XCMS 3 to avoid confusion with XCMS^2, leverages new R capabilities, packages, and objects to optimize memory usage. Unlike its predecessor, XCMS 3 does not load entire raw data files into memory but selectively loads the necessary portions, significantly reducing memory requirements. Moreover, the new object system enables users to save their object files and track the history of data processing methods applied to the object.

XCMS Online, introduced in 2012, offers a cloud-based and graphical system for utilizing XCMS. This web-based platform eliminates the need for local hardware resources and facilitates easy data sharing among users. It provides users with simple options to set up data processing workflows, accommodating various experiment types through different job types:

Single job: Analyzing a single file or a single class of data.

Pairwise job: Comparing two classes, such as knockout vs. wild-type experiments, with parametric or nonparametric statistical choices and options for unpaired or paired samples.

Multigroup job: Ideal for time series experiments involving multiple classes, employing ANOVA or Kruskal-Wallis statistical tests with post hoc analysis.

XCMS Guided Systems Biology

A systems-level analysis (**Figure 7.12**) is also useful for gaining important insights into the underlying biochemical mechanisms. XCMS Online offers a unique capability to project metabolomics data onto metabolic pathways and integrate it with transcriptomics and proteomics data, enabling a comprehensive understanding of cellular processes. Traditionally, projecting quantitative metabolic data onto metabolic networks requires prior identification of metabolites, a labor-intensive task relying on manual effort and expert curation. However, XCMS Online introduces an innovative approach called Mummichog leverages biochemical information to annotate features and map them onto metabolic pathways, bypassing the need for metabolite identification. This automated predictive pathway analysis method empowers scientists to interpret the data, formulate biochemical mechanisms, and generate hypotheses, which can then be confirmed through tandem MS analysis.

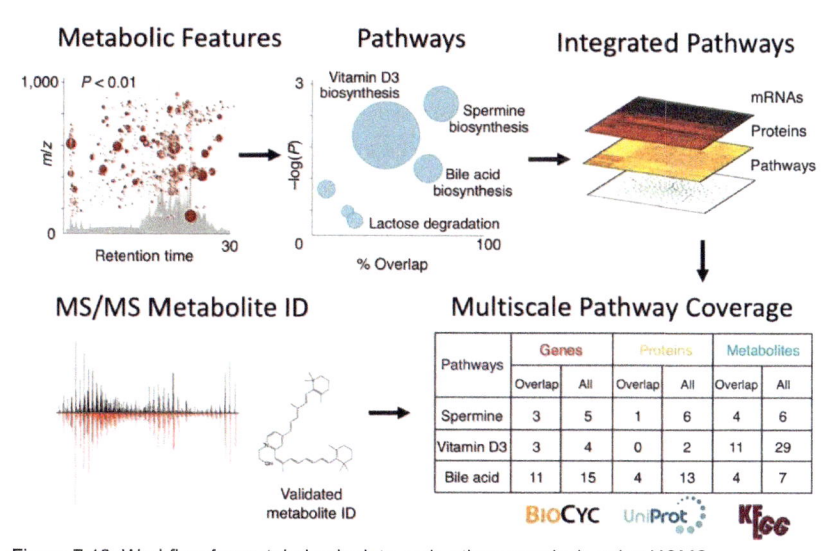

Figure 7.12. Workflow for metabolomic data and pathway analysis using XCMS. Statistically relevant metabolic features are generated from standard XCMS processing; these features undergo predictive pathway mapping. The pathway cloud plot shows dysregulated pathways (blue circles) with increasing statistical significance on the y- axis, metabolite overlap on the x-axis and total number of metabolites in the pathway represented by the radius. The multiscale pathway coverage table presents pathways overlapped across total metabolites, genes and proteins. MS/MS data helps confirm dysregulated pathways.

In the context of XCMS Online-guided systems biology, users can directly map their results onto metabolic networks without the hassle of transferring data between different applications. By uploading gene and protein data, users can overlay this information within the pathways, facilitating a multi-omics analysis. The results are presented in both tabular form and an interactive Pathway Cloud plot, which highlights dysregulated pathways and ranks them based on the overlap percentage with other omics data and statistical significance. Notably, XCMS Online offers access to over 7600 metabolic models for pathway analysis from BioCyc4 v19.5–20.0, providing researchers with a rich resource for comprehensive pathway exploration and interpretation. Through this integrated approach, XCMS Online empowers scientists to unravel intricate biological mechanisms and gain deeper insights into complex cellular processes.

XCMS MRM

For targeted metabolomics analysis using triple quadrupole MS configured for multiple reaction monitoring (MRM), XCMS-MRM (**Figure 7.13**) is the counterpart of XCMS Online. XCMS-MRM supports processing data from targeted metabolomics assays, automatically detecting transition peaks, integrating areas, and aligning data to minimize false peak integration. It offers quantification methods such as stable-isotope dilution, external calibration, and standard addition for acquiring absolute concentrations. XCMS-MRM also provides statistical analysis of concentration changes, quality control indicators, limits of detection, and unique graphical visualizations of the results.

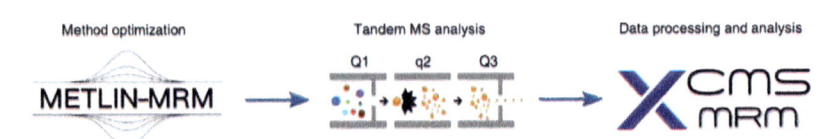

Figure 7.13. METLIN-MRM and XCMS-MRM are used in unison to examine targeted metabolomics data sets.

Overview

XCMS plays a valuable role in enabling LC-MS data acquisition and analysis for metabolomics research, catering to the evolving needs of the scientific community. This chapter has provided insights into how XCMS is used and its fundamental algorithms. As an R package, XCMS serves as a widely used and actively maintained software program for LC-MS data processing, enjoying widespread adoption within the community. On the other hand, XCMS Online extends the capabilities of XCMS by offering a comprehensive suite for metabolomics data analysis in a cloud-based environment.

With the support of high-performance dedicated servers at the Scripps Research Institute, XCMS Online delivers high processing speeds, making it easily accessible to a large user base. Researchers from diverse backgrounds can now leverage this resource to process their data. In summary, XCMS empowers scientists with the necessary tools to tackle the complexities of LC-MS data analysis in metabolomics, promoting collaboration and advancing our understanding of complex biological systems.

References

Smith, C. A., Want, E. J., O'Maille, G., Abagyan, R. & Siuzdak, G. XCMS: processing mass spectrometry data for metabolite profiling using nonlinear peak alignment, matching, and identification. *Anal. Chem.* **78**, 779–787 (2006).

Smith, C.A. et al. METLIN: a metabolite mass spectral database. *Ther. Drug Monit.* **27**, 747–751 (2005).

Xue, J., Guijas, C., Benton, H.P. *et al.* METLIN MS^2 molecular standards database: a broad chemical and biological resource. *Nat Methods* **17**, 953–954 (2020). https://doi.org/10.1038/s41592-020-0942-5

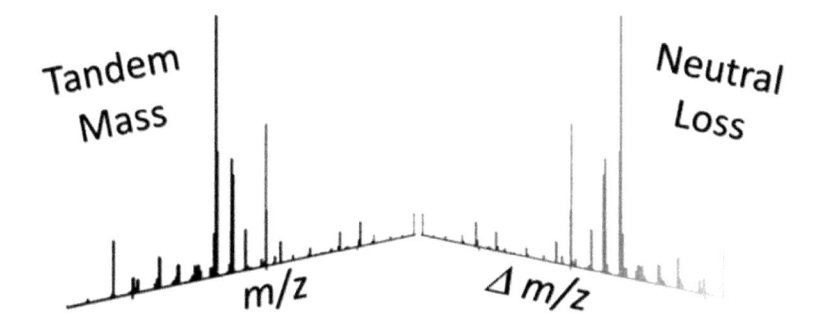

Chapter 8

METLIN

Perspective

Untargeted mass spectrometry metabolomics studies heavily rely on accurate databases to effectively identify metabolic features. By harnessing distinct fragmentation patterns, researchers can obtain valuable structural information for specific metabolites and molecular classes. In this chapter, we present the evolutionary journey of METLIN (**Figure 8.1**) as a comprehensive tandem mass spectrometry resource for small molecule analysis. Moreover, we discuss tools such as fragment similarity searching and neutral loss searching as they are employed to query METLIN, along with their corresponding workflows for the identification of uncharacterized molecular entities. Additionally, we describe other METLIN functionalities of a neutral loss database generated from MS/MS data, an extensive database of isotopic metabolites, an LC retention time database using reversed phase, a collision cross section database from ion mobility, and a multiple reaction monitoring database which aims to streamline the analysis of quantitative mass spectrometry data.

Figure 8.1. The logo that has come to represent METLIN is derived from its origins as a MS/MS database (right) of common endogenous metabolites. However, since 2005 it has grown to contain a vast array of molecules from over 350 chemical classes. METLIN is now widely used a a platform for identifying all types of molecules. It (like XCMS) originated with Colin Smith's efforts, however Winnie Uritboonthai (data acquisition) and Aries Aisporna (informatics) are the primary movers behind over 99% of METLIN's data content.

8. METLIN

Over the past two decades, the biological chemistry community has directed its attention towards endogenous metabolites, recognizing their significance in diagnostics, biomarker discovery, and activity. However, the utilization of mass spectrometry or metabolomics for these purposes posed several technical challenges. In particular, researchers worldwide faced significant obstacles in the characterization/identification of metabolites and other chemical entities. This chapter specifically focuses on the METLIN database and its associated tools. METLIN has evolved from its initial stage with only a few hundred metabolites to the current version, having a vast collection of experimental data on over 930,000 compounds as of January 2024 (**Figure 8.2**).

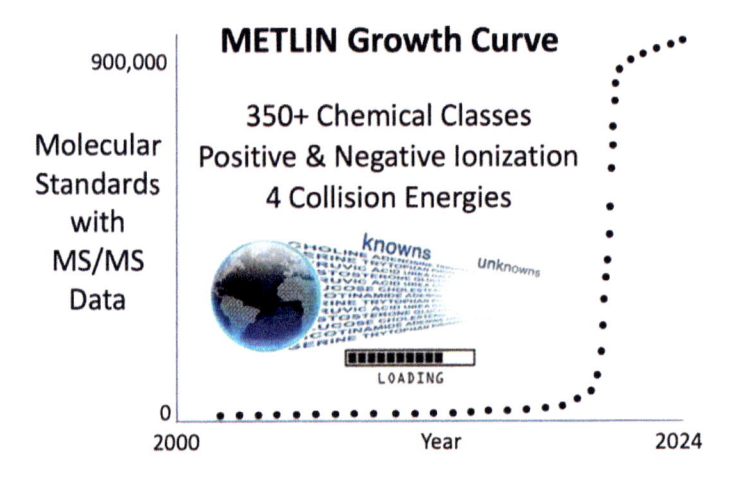

Figure 8.2. METLIN now encompasses experimental data on over 900,000 molecular standards. With success rate of over 80% on the standards, this translates into over 1.2 million molecules having been analyzed, each at four collision energies (0, 10, 20, 40 eV) and in positive and negative ionization modes.

METLIN-MS/MS

The METLIN-MS/MS database was originally developed with the primary objective of supporting metabolomic and lipidomic research across various disciplines, particularly aiding in the identification of metabolites in LC-MS-based metabolomics and bridging the gap between genomics, proteomics, and other "omics" sciences. The identification process for genes and proteins in genomics and proteomics experiments has been made possible through the predictive sequencing of nucleosides and amino acids, as well as the collaborative efforts of their respective communities. In contrast, the metabolome exhibits extensive chemical diversity and a vast array of molecular entities, making it challenging to predict fragmentation patterns (MS/MS) in MS-based experiments and thereby limiting the ability to make accurate identifications.

Since its inception in the early 2000s, METLIN-MS/MS has witnessed continuous growth in terms of the number of metabolites and MS/MS spectra, accompanied by the integration of new tools. After its public release in 2005, METLIN gradually expanded to over 10,000 metabolites with their corresponding MS/MS spectra in 2012, to where it stands now at over 900,000 (**Figure 8.2**). The database has various tools to facilitate and automate the identification of both known and unknown metabolites as will be described.

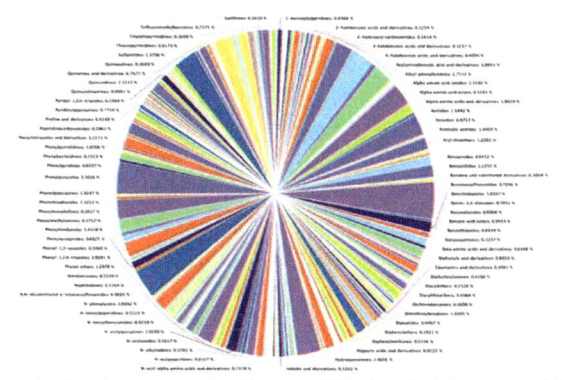

Figure 8.3. A crude visual representation the over 350 chemical classes and the experimental data on over 900,000 molecular standards within METLIN.

Alongside METLIN, several other academic, public, and private entities have developed metabolite/small molecule databases, categorized as either pathway-centric or compound-centric. METLIN however has adopted the approach of including any molecular entity as our lab believes that METLIN is a resource that is beyond metabolomics, and rather for all chemical entities (**Figure 8.3**).

METLIN-MS/MS also differs from other databases in its novel means of data collection and curation. All METLIN-MS/MS data is acquired

8. METLIN

at the Scripps Center for Metabolomics, adhering to strict protocols and employing standard reference materials that are commercially available or synthesized. Unlike other databases, METLIN does not use complex biological samples for reference fragmentation spectra to ensure minimal interference and optimal spectral quality. Over the years, METLIN has expanded its coverage to include molecular standards from a diverse range of molecular classes and biological or chemical origins. Therefore, METLIN offers a wide spectrum of small molecules, encompassing endogenous metabolites such as lipids, amino acids, nucleotides, and carbohydrates, toxicants, drugs, drug secondary metabolites, with molecules from over 350 chemical classes (**Figure 8.3**). The value of MS/MS data is demonstrated (**Figure 8.4**) where the redundancy of hits from "identifications" based solely on precursor mass is high, even with absolute accuracy (no error to four decimal places). Tandem MS/MS data greatly reduces the number of hits.

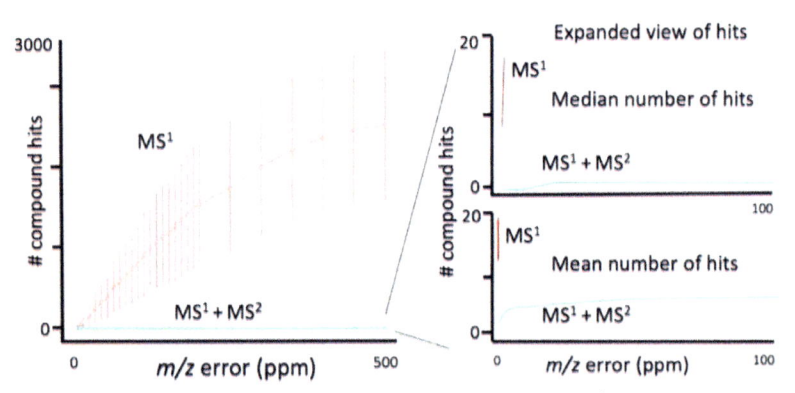

Figure 8.4. A sampling of METLIN hit rates (left) as a function of m/z error for MS1 data alone and combined MS1 and MS/MS data. The median and mean number of compound hits as a function of mass error created from the METLIN-MS/MS. The plot in red represents the number of hits based on precursor molecular weight as a function of error in part per million (ppm). The plot in blue represents the median and mean number of hits generated as a function of error, precursor molecular weight and tandem mass spectral fragments. An expanded view (right) of the hits generated from MS1 precursor ion data and MS/MS fragmentation data.

METLIN-MS/MS Basic Search Engines

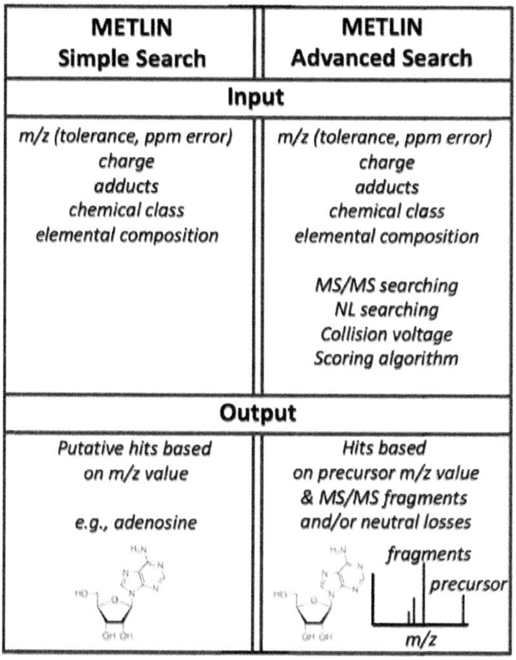

METLIN Simple Search	METLIN Advanced Search
\multicolumn{2}{c}{Input}
m/z (tolerance, ppm error) charge adducts chemical class elemental composition	m/z (tolerance, ppm error) charge adducts chemical class elemental composition MS/MS searching NL searching Collision voltage Scoring algorithm
\multicolumn{2}{c}{Output}
Putative hits based on m/z value e.g., adenosine	Hits based on precursor m/z value & MS/MS fragments and/or neutral losses fragments / precursor m/z

Figure 8.5. METLIN has multiple forms of searching with its "Simple Search" and "Advanced Search" capabilities using precursor m/z, chemical class, elemental composition, adducts, MS/MS and NL data.

In a traditional untargeted metabolomics workflow, dysregulated features, represented by specific m/z values and their corresponding retention time values, are typically searched against a database to identify putative metabolites based on accurate mass. METLIN offers multiple search options within its Simple and Advanced Search (**Figure 8.5**). The Simple Search enables the search of m/z values and neutral masses within a selected mass tolerance (**Figure 8.6**). METLIN allows the selection of several adducts in both positive and negative polarities. Similarly, the Batch Search provides the same functionalities as the Simple Search but allows the simultaneous search of multiple m/z values, facilitating the annotation of adducts and common losses associated with the same metabolite. This feature also helps differentiate ions originating from different molecules and link them to other putative m/z values.

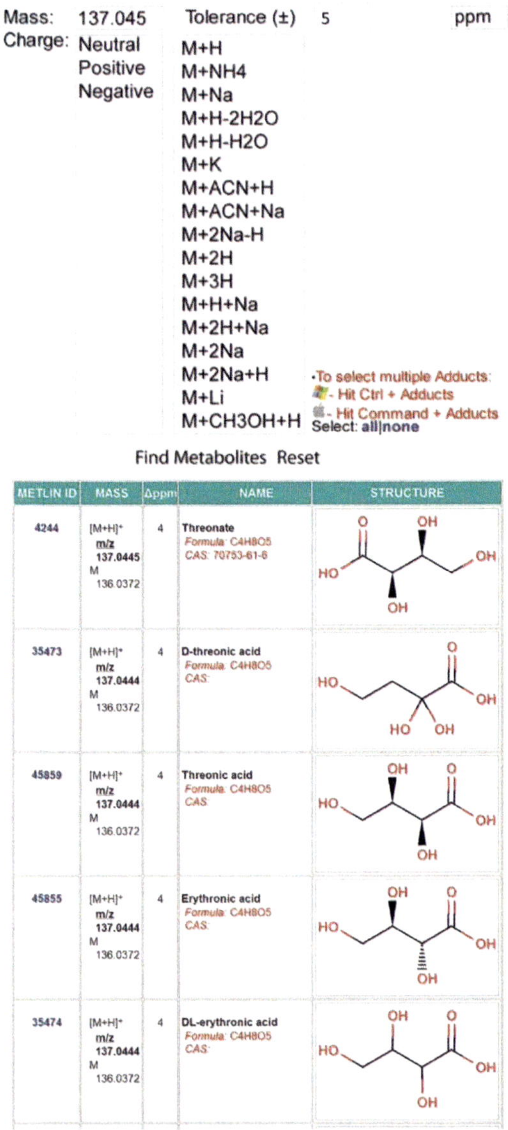

8. METLIN

Mass: 137.045	Tolerance (±) 5		ppm
Charge: Neutral	M+H		
Positive	M+NH4		
Negative	M+Na		
	M+H-2H2O		
	M+H-H2O		
	M+K		
	M+ACN+H		
	M+ACN+Na		
	M+2Na-H		
	M+2H		
	M+3H		
	M+H+Na		
	M+2H+Na		
	M+2Na		
	M+2Na+H	·To select multiple Adducts:	
	M+Li	- Hit Ctrl + Adducts	
	M+CH3OH+H	- Hit Command + Adducts Select: all|none	

Find Metabolites Reset

METLIN ID	MASS Δppm	NAME	STRUCTURE
4244	[M+H]+ 4 m/z 137.0445 M 136.0372	Threonate Formula: C4H8O5 CAS: 70753-61-6	
35473	[M+H]+ 4 m/z 137.0444 M 136.0372	D-threonic acid Formula: C4H8O5 CAS:	
45859	[M+H]+ 4 m/z 137.0444 M 136.0372	Threonic acid Formula: C4H8O5 CAS:	
45855	[M+H]+ 4 m/z 137.0444 M 136.0372	Erythronic acid Formula: C4H8O5 CAS:	
35474	[M+H]+ 4 m/z 137.0444 M 136.0372	DL-erythronic acid Formula: C4H8O5 CAS:	

Figure 8.6 METLIN Simple Search. The search panel with 137.045 inputed and M+H+ selected as the adduct. The multiple hits demonstrate that m/z alone cannot for identification.

On the other hand, the Advanced Search allows users to search based on additional information such as molecular formula, compound names, SMILES, KEGG, CAS, and MID numbers. These search options generate a list of metabolite names (or molecular formulas) that could potentially correspond to the dysregulated feature of interest. However, due to the limited elemental diversity in biomolecules (C, H, N, O, P, and S), this list can contain numerous putative metabolite identifications, ranging from tens to hundreds.

METLIN Identification via MS/MS Matching

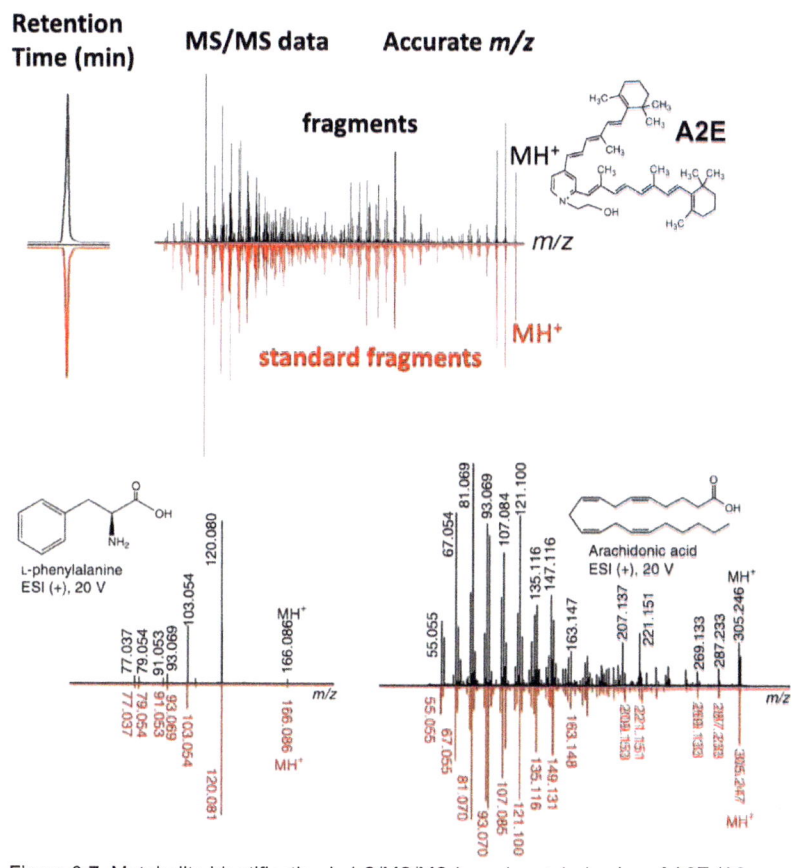

Figure 8.7. Metabolite identification in LC/MS/MS-based metabolomics of A2E (A2-ethanolamine), phenylalanine, and arachidonic acid. The identification includes accurate mass of the compound compared to the theoretical mass based on the compound's molecular formula. Retention time is also used. Most notably, assignment is highly weighted on MS/MS data.

8. METLIN

To refine the list of putative identifications obtained from *m/z* precursor ion searching, retention time, experimental MS/MS spectra which are compared against respective spectral libraries fragmentation patterns (**Figure 8.7**). The MS/MS Spectrum Match Search tool allows for autonomous metabolite identification. Users upload the fragmentation profile as a table of *m/z* values and intensities, along with the precursor mass within a specific tolerance, collision energy, and polarity.

This tool then searches, compares, and scores the similarity of the experimental spectra with the reference spectra in the library for all metabolites within the selected mass tolerance, utilizing a modified X-Rank similarity algorithm. Undoubtedly, the collection of over 1,000,000 molecular standards with corresponding MS/MS spectra is METLIN's most valuable contribution to metabolomics research. These spectra have been acquired at four different collision energies (0, 10, 20, and 40 eV). In METLIN this is performed in both positive and negative ionization modes.

In the realm of identifying unknown metabolites, spectral libraries serve a dual purpose. As previously mentioned, these libraries primarily facilitate identification by comparing experimental spectra with reference spectra, enabling identification up to level 2 according to the Metabolomics Standards Initiative. However, despite significant efforts to expand their populations, no library can claim completeness due to the vast number of metabolites and the diverse range of chemistries. In fact, the exact count of metabolites in nature (animals, plants, eukaryotes, prokaryotes and fungi) remains a topic of debate, although estimates in the millions are not exaggerated. Such numbers greatly surpass the approximately 20,000 genes and proteins.

Therefore, spectral libraries serve a second crucial purpose: assisting in the identification and characterization of both known metabolites lacking MS/MS spectra and unknown metabolites that have not been previously documented in any library or resource. To fulfill this purpose, two tools, namely Fragment Similarity Search and Neutral Loss Search, have been developed and continuously enhanced over the past 11 years. These tools capitalize on the vast amount of spectral information available and exploit the similarities in dissociation routes among compounds with related structures and chemical moieties (**Figure 8.8**).

In this context, we present an illustrative example showcasing the utilization of these tools for the identification of unknown metabolites. Following unsuccessful attempts to identify a specific metabolic feature through accurate mass and MS/MS spectra matching, we turned to the Fragment Similarity Search and Neutral Loss Search tools to obtain structural information and thereby gain insights into the molecular identity. The subsequent steps involved:

We utilized the Fragment Similarity Search tool to search for selected fragments from the MS/MS spectrum of the unknown metabolite (**Figure 8.8**). These fragments, namely 179.03, 235.10, 299.09, and 355.15, were chosen based on their higher intensity, which indicates their potential to provide valuable structural information. The analysis yielded multiple hits in the METLIN database, with xanthohumol matching all four fragments, suggesting the presence of a metabolite containing a similar chemical structure or moieties.

Assuming that the presence of xanthohumol represents a part of the molecule of interest, calculating the mass difference between the precursor ion and this potential glucuronide fragment (531.18 - 355.15 = 179.03), we employed the Neutral Loss Search tool within a selected ppm window to search for fragments with a mass of 179.03. Most of the Neutral Loss Search results indicated molecules containing glucuronide, suggesting that glucuronide may be the second component of the unknown metabolite.

8. METLIN

METLIN Similarity Searching to Characterize Unknowns

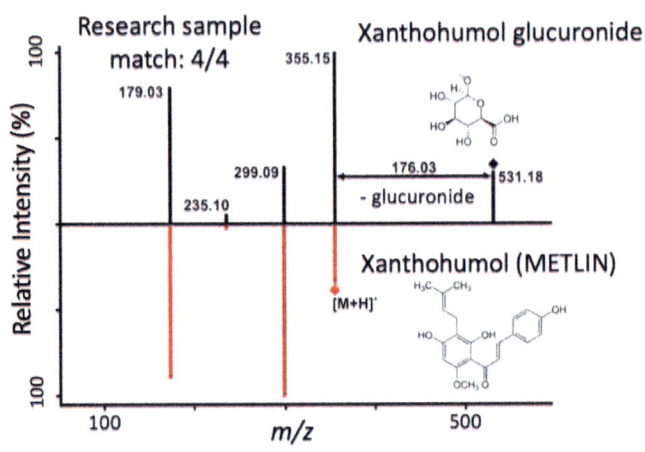

Figure 8.8. Fragment Similarity Search facilitates the identification of unknown metabolites where no MS/MS spectral data are available. The fragments of an unknown metabolite were searched against METLIN and four fragments were found to match with xanthohumol. The comparison between experimental and library MS/MS spectra implies high structural similarities and further analysis of neutral loss was consistent with xanthohumol glucuronide.

METLIN - NL

Neutral loss (NL) spectral data presents a mirror of MS/MS data (**Figure 8.9**) and is a valuable yet largely untapped resource for molecular discovery and similarity analysis. Tandem mass spectrometry (MS/MS) data is effective for the identification of known molecules and the putative identification of novel, previously uncharacterized molecules (unknowns). Yet, MS/MS data alone is limited in characterizing structurally related molecules. To facilitate unknown identification and complement the METLIN-MS/MS fragment ion database for characterizing structurally related molecules, we have created a MS/MS to NL converter to generate METLIN-NL. Mirroring MS/MS data to generate NL spectral data offers a unique dimension for chemical and metabolite structure characterization. METLIN-NL (NLintensity vs. delta *m/z*) spectra was generated by calculating the difference between the precursor and fragment ions with NLintensity based on the original fragment ion intensities.

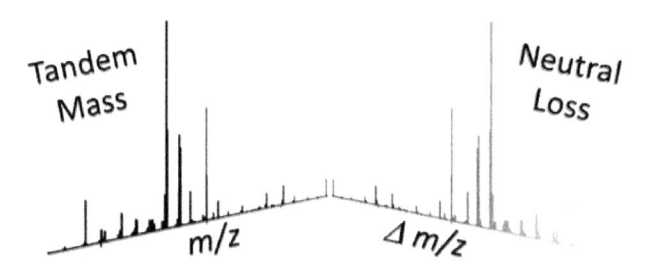

Figure 8.9. Tandem MS and NL spectra mirror each other and thus NL offers an alternative dimension in characterizing molecules.

To test the utility of METLIN-NL we examined two structurally related oxylipins. Oxylipins represent a class of highly active lipid metabolites ubiquitous in humans and plants, and specifically, the phytoprostanes (PhytoPs) class of oxylipins resemble prostaglandin-like compounds that are found in seeds and vegetable oils derived from oxidative cyclization of α-linolenic acid. Since PhytoPs are a class of highly structurally related oxylipins and are suspected to have additional unidentified analogs, we chose them to demonstrate the utility of METLIN-NL.

Tandem MS and NL data were recently generated on a set of PhytoPs, including the structural analogs 16-B1-PhytoP and 16-keto 16-B1-PhytoP (**Figure 8.10**). When trying to extrapolate/correlate the observed tandem MS spectra of the two PhytoPs, classic similarity searching was of very limited value providing only one overlapping ion. This exemplifies that two structurally very similar molecules can yield highly different MS/MS spectra limiting similarity searching possibilities and thereby severely impacting the usefulness of this approach for the identification of chemically closely related substances. However NL similarity analysis yielded multiple overlapping NLs (**Figure 8.10**) and helped to easily correlate the two molecules.

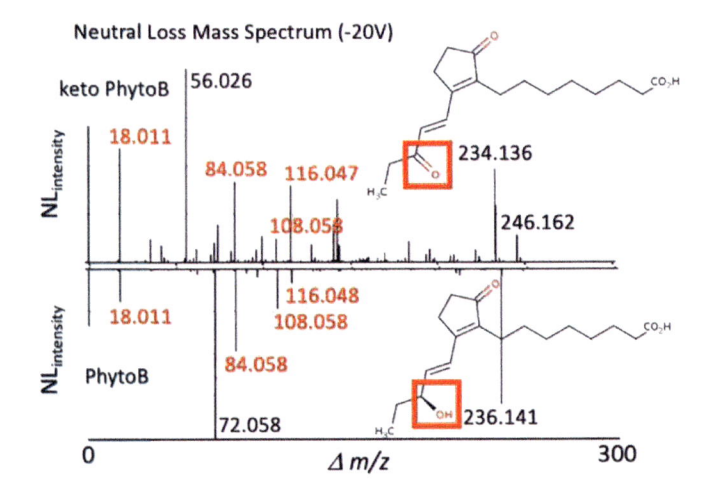

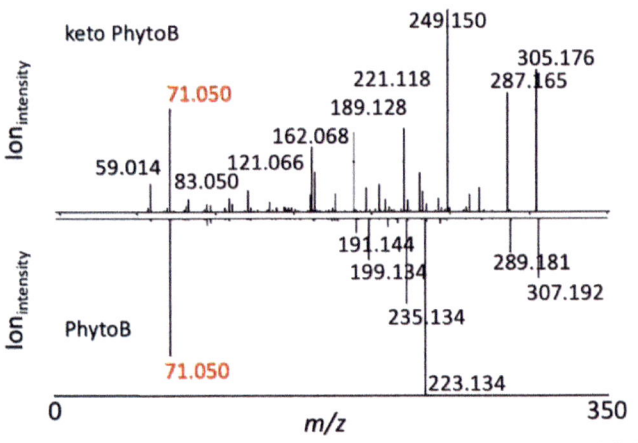

Figure 8.10 NL and MS/MS data on two very similar oxylipins (16 keto 16-B1-PhytoP and 16-B1-PhytoP). The NL spectra with the neutral loss data shows a high degree of similarity while the MS/MS data has only one significant overlapping fragment. (in red).

METLIN - Iso

METLIN-Iso in parallel with the development of METLIN has been introduced to leverage the expanding number of analytical standards available in METLIN. Created in 2014, METLIN-Iso serves as a database for isotope incorporated metabolites. With a user-friendly interface akin to METLIN, METLIN-Iso provides accurate mass information for all computed isotopologs present in METLIN, encompassing compounds with varying numbers of isotope-labeled atoms and, consequently, different *m/z* values. The METLIN-Iso Search feature includes the stable isotopes in labeling experiments, such as 13C, 15N, 2H, and 18O, can be distinguished through accurate mass measurements, further analysis of their MS/MS spectra is crucial for determining the precise position of the isotopic label within the same isotopologs. This feature can be useful for investigating metabolic pathways. To address this need, METLIN-Iso encompasses the MS/MS spectra of hundreds of isotopomers (i.e., the same isotopologs with different locations of labeled atoms). The collection of spectra aids in the tracking of isotopic labels and provides comprehensive information regarding the de novo synthesis of metabolites within specific pathways. While the primary application of METLIN-Iso revolves around the analysis of stable isotopes in labeling experiments, it also offers a wealth of insights into metabolic pathways and their intricate dynamics.

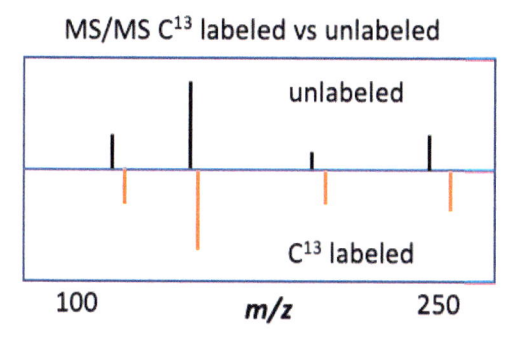

MS/MS C^{13} labeled vs unlabeled

Figure 8.11 Systematic generation of MS/MS spectra of uniformly labeled metabolites in isoMETLIN. C13-labeled and labeled glucose, producing uniformly labeled metabolites. The MS/MS spectra help structure elucidation by using the number of carbons that are labeled in each fragment.

It is worth noting that because of the cost of isotopically labeled molecules and the very limited collection of available molecules, METLIN-Iso is not nearly as extensive as METLIN-MS/MS and contains a few thousands of molecules.

Similar to METLIN, METLIN-Iso has fragmentation spectra acquired from qToF instruments at various collision energies using authentic isotope-labeled standards. However, one major challenge in populating the spectral database with MS/MS spectra of metabolite isotopologs is the exponential increase in the number of isotopomers with molecular weight (number of atoms), and the limited availability of commercially labeled isotopomers.

Fully labeled MS/MS spectra provide valuable structural information for metabolite identification. By utilizing the mass differences between analogous fragments of isotopologs, it becomes possible to determine the number of carbon atoms in each fragment (**Figure 8.11**). This information is particularly useful for the identification of metabolites where MS/MS data is not available.

METLIN - RT

The METLIN-RT library comprises data on over 80,000 molecular standards, it was created since reversed phase (RP) liquid chromatography coupled to mass spectrometry (LC–MS) is one of the more widely used chromatography approaches. One of the functions of METLIN-RT was to help train retention time prediction algorithms using machine learning (**Figure 8.12**).

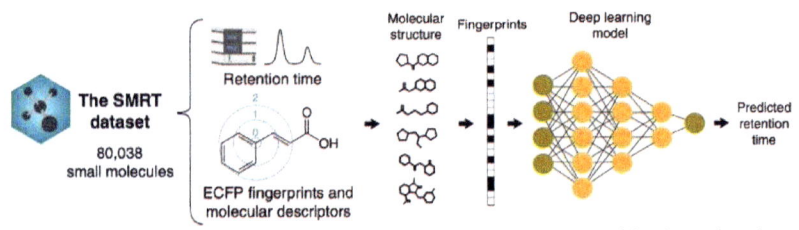

Figure 8.12 Composition of the "SMRT" 80,000 dataset and structure of the deep-learning model that was used to create a retention time prediction model.

METLIN - MRM

METLIN-MRM comprises small-molecule transitions for multiple-reaction monitoring (MRM) generated from our METLIN-MS/MS database. It was developed to streamline absolute quantitation, typically performed using triple quadrupole (QqQ) mass spectrometers configured to monitor specific precursor-product ion transitions. Typically, determining transitions for different target molecules requires optimization with pure standard materials. The METLIN-MRM library provides traditional experimentally optimized transitions and computationally optimized experimental transitions derived from the METLIN-MS/MS data. Experimentally optimized transitions were acquired for over 1000 molecules in both positive and negative modes following established protocols. These transitions were optimized to achieve maximum sensitivity and selectivity (**Figure 8.13**).

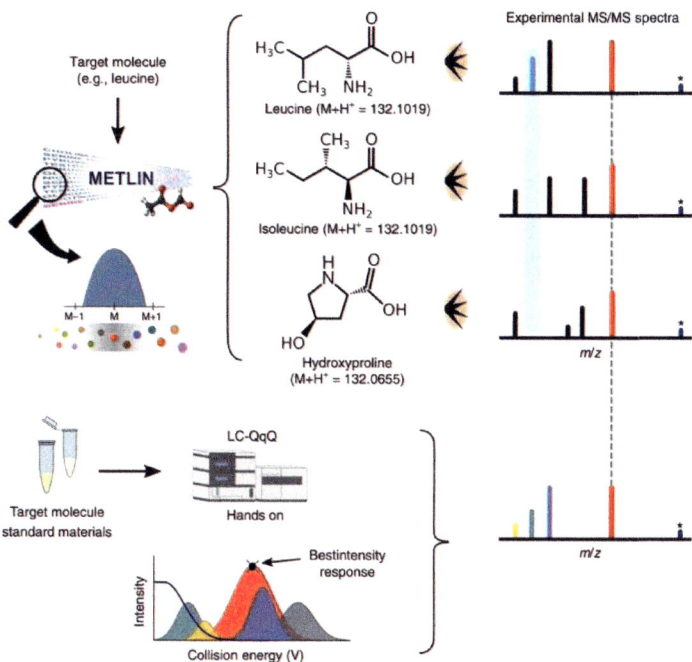

Figure 8.13. METLIN-MRM. In the METLIN-MRM ranking system, all molecules in the METLIN library with a precursor within a ±0.7-Da window of the target molecule are compared (for example, leucine is compared with isoleucine and hydroxyproline), and candidate transitions are selected on the basis of their fragment selectivity (blue fragment). Experimental QqQ transitions are optimized by using standard materials and prioritizing high-intensity fragments (sensitivity), which might yield nonspecific fragments (for example, red fragment) and thus misidentifications.

In addition to experimentally acquired data, transitions for more than 14,000 and 4700 molecules in positive and negative modes, respectively, were computationally optimized using the METLIN-MS/MS data. The spectral library was obtained by acquiring spectra at different collision energies on a qToF instrument, and an empirical ranking algorithm was developed to select transitions based on the selectivity (uniqueness of a product fragment for a given molecule) of empirical MS/MS fragments. This approach allows for high-throughput quantitation analysis, eliminating the need to optimize transitions with standard reference materials. It also reduces errors caused by interfering molecules, as less likely masked transitions are selected. More detailed information about the algorithms and selection process can be found in the main published work.

METLIN-CCS

Ion mobility spectrometry (IMS) has emerged as an important separation technique for small molecule characterization in biochemical research. The METLIN-CCS database, which includes collision cross section (CCS) values and IMS data for over 27,000 molecular standards representing 79 chemical classes. METLIN-CCS provides CCS values measured in triplicate in both positive and negative ionization modes, yielding multiple ion types (e.g., M+H+, M+Na+, M+NH4+, [M-H]-, M+Cl-, and M+TFA-) With over 185,000 CCS values, the METLIN-CCS database is a unique resource for small molecule characterization and IMS-based machine learning.

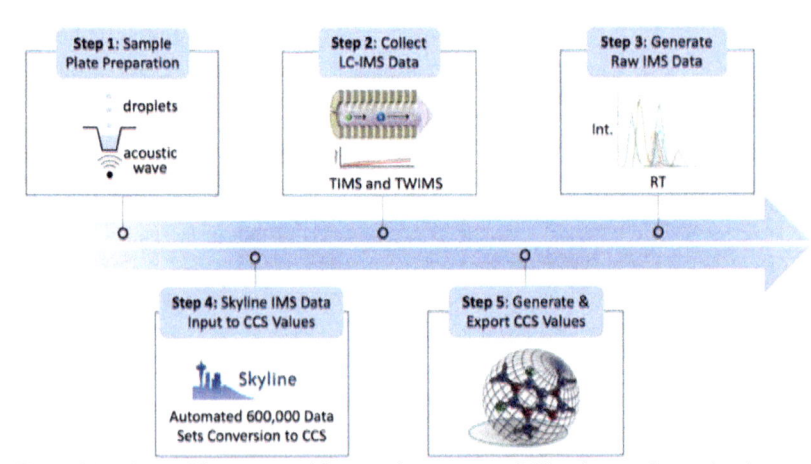

Figure 8.14. General five step workflow used to generate CCS values on the molecular standards.

The METLIN-IMS database was designed for three purposes: (1) create a downloadable resource containing thousands of CCS values, (2) provide a basis set of data to train machine learning models, and (3) explore aggregation properties of each standard to provide information on how molecular ions behave in the source and throughout the instrument. The METLIN-CCS database was generated in a five-step process (**Figure 8.14**) by analyzing molecular standards in triplicate and in both positive and negative ionization mode. The standards represented 79 different molecular classes. Ultimately, ~185,000 CCS values matching ~62,000 unique molecular species and ~28,000 standards were retained for the METLIN-IMS database.

Overview

METLIN encompasses experimental data on over 900,000 molecular standards spanning 350 molecular classes. The library has been continuously expanded without bias towards any specific class or type of compounds. It includes endogenous metabolites from eukarya (protozoa, yeast, plants, animals, and fungi), archaea, and bacteria domains, modified metabolites, synthetic drugs, and toxicants. The growth of the MS/MS database has recently been exponential, now at over 931,000 in January 2024 with data in both positive and negative ion modes at multiple collision energies. This growth trend is expected to continue for as long as I can find compounds.

In conclusion, this chapter has highlighted the evolution of METLIN from its inception and discussed its tools for metabolite identification. METLIN has successfully adapted and developed its tools to not only aid in the identification of known compounds, with or without MS/MS spectra, but also to enable the discovery of unknown compounds. Additionally, the METLIN family has expanded to include all of the following databases, ordered by size.

References

Smith, C.A. et al. METLIN: a metabolite mass spectral database. *Ther. Drug Monit.* **27**, 747–751 (2005).

Smith, C. A., Want, E. J., O'Maille, G., Abagyan, R. & Siuzdak, G. XCMS: processing mass spectrometry data for metabolite profiling using nonlinear peak alignment, matching, and identification. *Anal. Chem.* **78**, 779–787 (2006).

Xue, J., Guijas, C., Benton, H.P. *et al.* METLIN MS^2 molecular standards database: a broad chemical and biological resource. *Nat Methods* **17**, 953–954 (2020). https://doi.org/10.1038/s41592-020-0942-5

Chapter 9

Metabolite Discovery

Perspective

Metabolite identification is a significant challenge and also presents a valuable opportunity in the field of biochemistry. The comprehensive characterization and quantification of metabolites in living organisms across the centuries (**Figure 9.1**) have laid the foundation for a vast biochemical knowledgebase, contributing to the resurgence of metabolism research in the 21st century. However, the identification of newly observed metabolites continues to be a persistent obstacle.

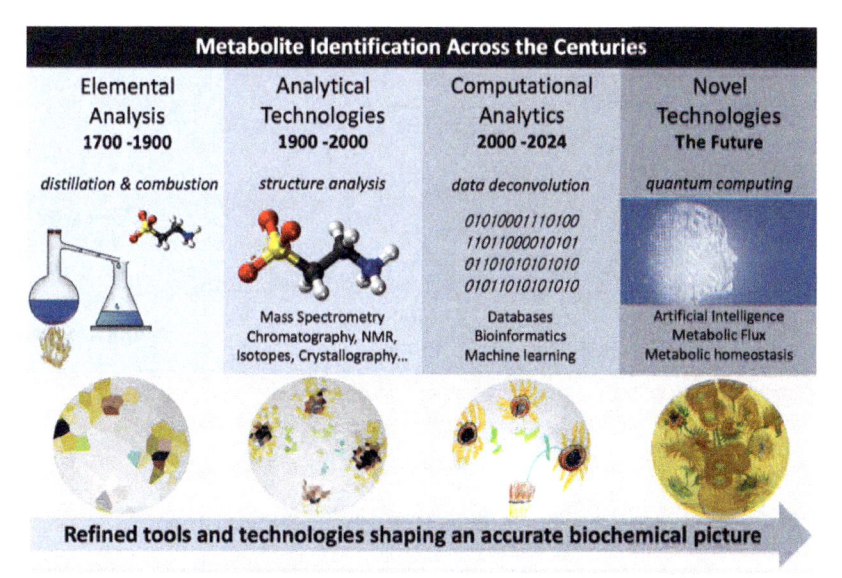

Figure 9.1. A broad depiction of the evolution of metabolic sciences and metabolite identification across the centuries as depicted by technology and (below) an art metaphor to van Gogh (credit Paul Giera).

Crystallography and NMR spectroscopy have played crucial roles in advancing our understanding of metabolism. Nevertheless, their applicability in unraveling the intricate details of metabolism's fine structure has been limited by the need for sufficient and highly pure materials. Mass spectrometry, particularly when combined with high-performance separation techniques and emerging informatics, AI, and database solutions, has emerged as a pivotal technology in metabolite identification (**Figure 9.2**).

In this chapter, the historical approaches that have been employed to tackle the metabolite identification conundrum will be briefly covered with an emphasis on role of mass spectrometry in advancing our understanding of metabolism. Additionally, we discuss how metabolomics is evolving in response to these challenges and how the integration of AI technologies is shaping the future of metabolite characterization.

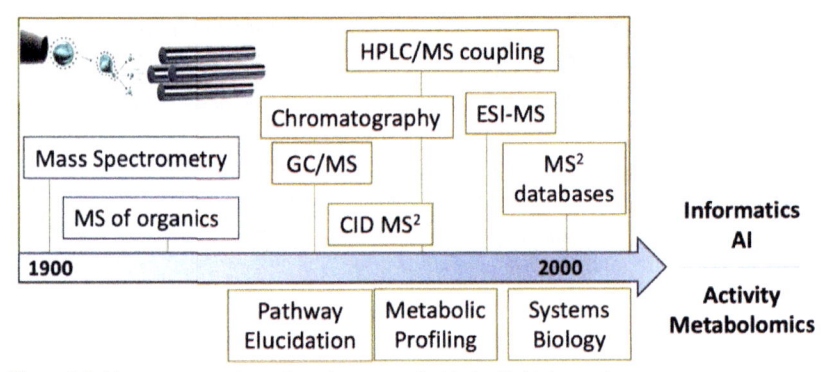

Figure 9.2. How mass spectrometry advances coincided with biology advances.

Elemental Composition Structure Determination

The elemental composition determinations of small organic molecules, including urea, lactic acid, citric acid, and oxalic acid (**Figure 9.1**), originated from the pioneering analytical techniques developed by Boerhaave and Lavoisier in the 1700s. Subsequently, Gay-Lussac and Thenard made significant improvements to these techniques in the early 1800s. Initially, their approach involved

9. Metabolite Discovery

analyzing animal and food products known to contain high concentrations of specific molecules. For instance, citric acid was sourced from lemons, and lactic acid was extracted from fermented milk. The subsequent steps included the separation and purification of the constituents through distillation and crystallization. Atomic weights were then derived through combustion analysis. Although these chemical formulas provided valuable insights, they only served as a basis for structural hypotheses.

During the nineteenth century, significant progress was made in determining the molecular formulas of various metabolites. Justus von Liebig's book "Animal Chemistry" played a crucial role in establishing our understanding of metabolic reactions, even though these reactions had not been observed in vivo at that time. Liebig's work laid the foundation for studying the inter-conversions of simple organic molecules within cells. The use of radioactive isotopes in the twentieth century provided experimental evidence for the predicted metabolic reactions. Techniques such as crystallography and physico-chemical analysis, including combustion analysis and boiling/melting point determination, contributed to the characterization of substances. Although these efforts didn't always yield confirmed metabolite structures, they were significant intellectual achievements considering the available analytical technologies. Today, many of these techniques and properties are still taught in chemistry and pharmacy studies. Notable examples of reagents developed during the 1800s include the Tollens test and the Marquis reaction. The determination of cholesterol's chemical structure spanned over two centuries, with various researchers making important contributions. The total synthesis of cholesterol was achieved in 1951/52, nearly 200 years after its initial discovery.

In the 1900s, significant technological advancements revolutionized biochemical research. Developments such as x-ray crystallography, nuclear reactors for artificial radioisotopes, and scintillation spectrometers led to exponential growth in the field. Hans Krebs discovered the TCA cycle, acetyl-CoA, glycolysis, and steroid biosynthesis. Chromatographic techniques like gas chromatography and high-pressure liquid chromatography emerged. By 1957, biosynthetic pathways for various biological molecules were elucidated. Nuclear magnetic resonance (NMR) and mass spectrometry (MS) emerged as powerful tools for metabolomics. The concept of individual biochemical profiles was introduced by Roger J. Williams. GC/MS methods were developed to monitor metabolites. Electrospray ionization (ESI) and LC-ESI MS paved the way for untargeted metabolomics. Genomics and proteomics gained attention, driving the establishment of systems biology. Bioinformatics and data mining pipelines were developed for large datasets. Despite these advances, the focus on genomics and proteomics led to the notion that metabolite investigation was mature. Textbook metabolic pathways became the primary target for biomarker discovery, overshadowing efforts to understand the metabolic machine comprehensively.

Analytics, Databases & Bioinformatics in Identification

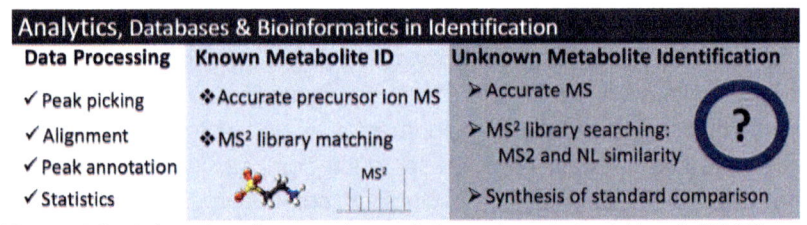

Data Processing	Known Metabolite ID	Unknown Metabolite Identification
✓ Peak picking	❖ Accurate precursor ion MS	➢ Accurate MS
✓ Alignment	❖ MS² library matching	➢ MS² library searching: MS2 and NL similarity
✓ Peak annotation	MS²	
✓ Statistics		➢ Synthesis of standard comparison

Figure 9.3. Peak detection, alignment, deconvolution, and spectral matching via MS/MS databases of standards, and statistical assessment of metabolomic data followed by known and unknown metabolite identification. The timeline for unknown identification can vary from days to years depending on the complexity of the chemical structure. Also, it depends on the complexity of the synthesis and the amount of biological material available. Unknown characterization is not an easy process.

In the 21st century, advances in genomics, proteomics, and bioinformatics allowed for large-scale analysis of DNA/RNA and proteins. Metabolomics, which focuses on metabolites, lagged behind due to metabolite's unique characteristics. However, in the past two decades (**Figure 9.3 & 9.4**), improved LC-MS/MS techniques coupled with bioinformatics enabled comprehensive metabolome-wide investigations. Analytical hardware improvements increased metabolome coverage despite challenges posed by the wide concentration range of metabolites. The availability of comprehensive databases for metabolite identification, although still limited, is growing rapidly. Artificial intelligence (AI) and machine learning (ML) have emerged as powerful tools in metabolomics, aiding in metabolite identification, spectral analysis, and prediction.

9. Metabolite Discovery

Deep learning (DL) and natural language processing (NLP) offer additional avenues for data interpretation and knowledge extraction from the scientific literature. However, identifying complex chemical structures from limited numeric variables remains a challenge, and empirical data and confirmation through orthogonal techniques like NMR are still necessary for unequivocal identification.

Emerging technologies such as cryogenic electron microscopy (cryo-EM) and novel fragmentation techniques in mass spectrometry (MS) hold promise for metabolite identification. Separation technologies like supercritical fluid chromatography and advanced liquid chromatography (LC) are advancing metabolite identification. Nuclear magnetic resonance (NMR) remains the gold standard for metabolite identification, especially with sufficient and pure material. Analytical advancements, bioinformatic tools, and spectral databases have shifted biology towards unbiased global investigations. The presence of known and unknown molecules correlated with specific biological conditions has led to the discovery of a "metabolic black matter" that is yet to be fully understood. Stable isotope labeling provide insight into biosynthetic origin and fate.

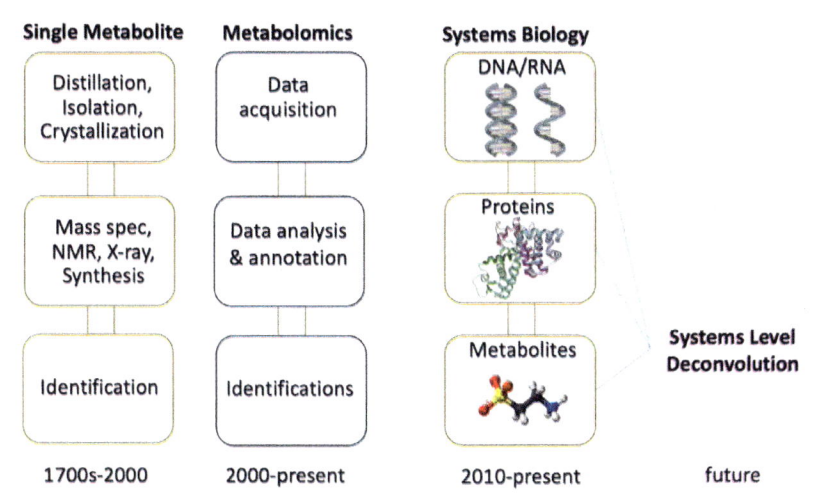

Figure 9.4. Multiple centuries passed as we struggled to identify individual metabolites, now in the course of decades the process of identifying metabolites has become more routine. The ongoing challenge is in identifying all metabolites and deconvolving their systems level impact.

Identifying Metabolites/Lipids (Oleamide): Sleep

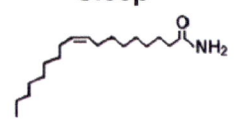

In the early 1990s I performed LC/MS experiments designed to identify molecules in cerebral spinal fluid associated with sleep, "Chemical Characterization of a Family of Brain Lipids That Induce Sleep" Science 1995 & PNAS 1994 (**Figure 9.5**). In these studies, we aimed to identify and characterize compounds that induce sleep by extracting metabolites and lipids from the CSF of sleep-deprived felines. This was the first LC/MS-based metabolomics experiments, although it was not called that at the time.

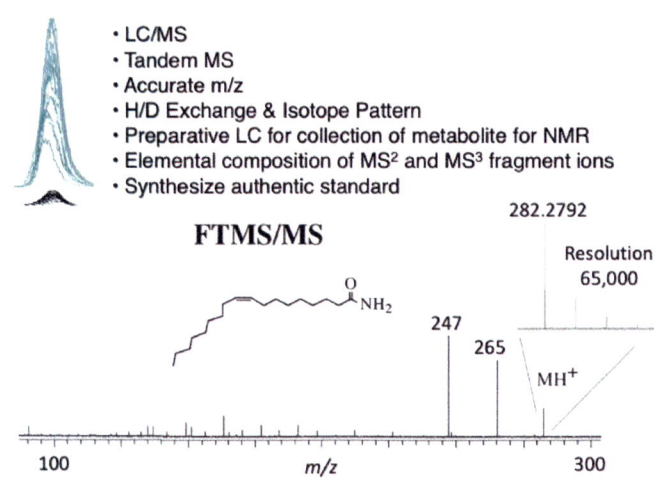

- LC/MS
- Tandem MS
- Accurate m/z
- H/D Exchange & Isotope Pattern
- Preparative LC for collection of metabolite for NMR
- Elemental composition of MS^2 and MS^3 fragment ions
- Synthesize authentic standard

FTMS/MS

Figure 9.5. Data that was acquired during and since the original sleep study that helped identify oleamide.

Through these experiments, we observed a family of related fatty acid amide lipids, one in particular stood out with an m/z of 282.279 was cis-9,10-octadecenoamide, which is commonly referred to as oleamide. Probably the most significant outcome of this effort was 1) oleamide acted as a sleep-inducing compound and, more importantly 2) the recognition of the significant time it took to manually analyze the data and identify the metabolites. This served as the springboard behind the creation of XCMS and METLIN.

9. Metabolite Discovery

Identifying Lipid Neuroprotectin D1: Stem Cells and the "Plastic Metabolome"

This study (**Figure 9.6**, Nature Chemical Biology 2010) explored the metabolomic profile and redox regulation in embryonic stem cells (ESCs) during differentiation. By employing mass spectrometry-based metabolomics, Oscar Yanes identified a distinct metabolic signature characterized by highly unsaturated metabolites in ESCs, which decreased upon differentiation.

In his own words "Our results reveal that the pluripotent properties of stem cells can be derived by the low oxidative state of their metabolomes. The presence of specific highly unsaturated endogenous metabolites confers the necessary chemical plasticity to differentiate into new structural and more oxidized and hydrogenated metabolites detected in mature populations. We postulate that the existence of such a "plastic metabolome" is promoted and maintained by the known hypoxic microenvironments where most of the stem cells reside in living organisms."

This was demonstrated in the redox status, as indicated by the ratio of reduced and oxidized glutathione and ascorbic acid levels, as well as the rich population of unsaturated lipids in stem cells (**Figure 9.6**). This data was meticulously generated using an FT-ICR-MS system that allowed for the careful characterization of elemental composition and thus the hydrogen deficiency index (HDI) which allowed him to discern the high level of unsaturation in stem versus mature cells.

One of the interesting molecules that came out of this was neuroprotecting D1. Neuroprotectin D1 (NPD1) is a specialized lipid mediator derived from an omega-3 fatty acid. NPD1 was found to significantly promote neuronal differentiation. The researchers supplemented neuronal differentiation media with NPD1 and observed a remarkable increase of over 1500% in the number of tubulin positive neurons, indicating enhanced neuronal differentiation. This finding suggests that NPD1 plays a crucial role in driving the differentiation of stem cells into neurons.

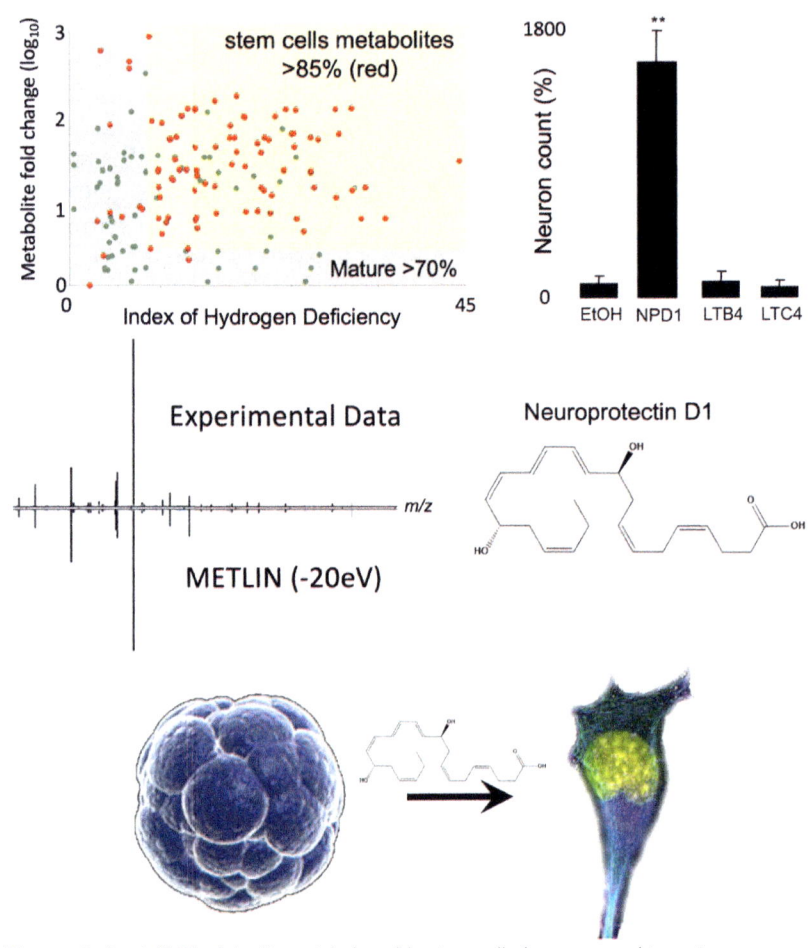

Figure 9.6. (top left) The "plastic metabolome" in stem cells (as compared to mature cells) represented by the index of hydrogen deficiency. One particular metabolite (NPD1, top right and bottom) drives stem cell differentiation 1500% more than controls. The MS/MS data that was acquired since the original study on NPD1 and its comparison to data in METLIN (-20eV) used for the identification.

Interestingly, NPD1 specifically exerted its neurogenic effects, while other pro-inflammatory eicosanoids, such as leukotriene B4 and leukotriene C4, had no significant impact on neuronal differentiation. This highlights the specific role of NPD1 in promoting neurogenesis. The study also conducted preliminary experiments on human ESCs, which demonstrated that the neurogenic activity of NPD1 is conserved across different species. Overall, the findings of this study suggest that endogenous metabolites, particularly NPD1, play a critical role in regulating the differentiation of stem cells into neurons.

Identifying Metabolite 3-Indole Propionic Acid: Gut Microbiota

The human gut microbiome, consisting of trillions of bacteria in the distal intestinal tract, plays a crucial role in human health. However, the specific biochemical effects of these bacteria on the host remain poorly understood. In this PNAS 2009 study (**Figure 9.7**), a comprehensive metabolomics analysis using mass spectrometry was to investigate the influence of the gut microbiome on mammalian blood metabolites. By comparing plasma extracts from germ-free mice (lacking gut bacteria) to those from conventional (wild type) mice, we identified that roughly 10% of the plasma metabolome was significantly altered. A level of alteration I had not previously experienced in our metabolomic studies.

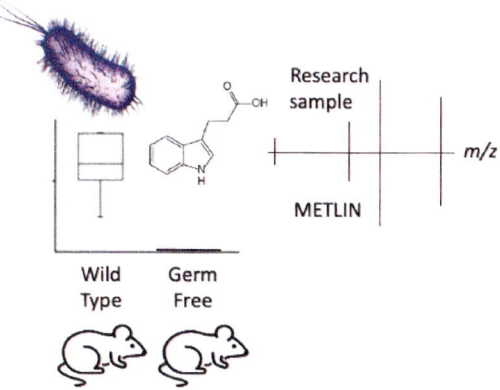

Figure 9.7. One of the more interesting metabolites was 3-indole propionic acid (3-IPA). 3-IPA was determined to come solely from the bacteria clostridium sporanges and has since has had over 2000 papers published on its activities.

One particular group of metabolites affected by the gut microbiome was indole-containing compounds. For example, we observed major alterations in the levels of tryptophan, N-acetyl tryptophan, and serotonin in conventional mice compared to germ-free mice. While the conversion of tryptophan to indole by specific gut bacteria explained the decrease in tryptophan levels, the increased serotonin levels in conventional mice suggested alternative, previously unobserved, mechanisms. Most interesting was the observation of 3-indole propionic acid (3-IPA), a potent antioxidant, exclusively in the plasma of conventional mice. While the shear impact of the gut microbiome had on the plasma metabolome was quite surprising in 2009, equally as interesting was the observation of gut specific metabolites in blood. 3-IPA has since gained considerable interest with thousands of papers being published on its unique properties, namely:

- 3-IPA is produced by the gut bacteria Clostridium sporogenes via the digestion of tryptophan.

- Neuroprotective properties including antioxidant and anti-inflammatory effects in the brain.

- Impact on intestinal function with evidence suggesting that IPA could influence the integrity of the intestinal barrier.

- IPA has been investigated in the context of several diseases, including neurological, inflammatory, and metabolic diseases.

Identifying Lipid N,N-Dimethylsphingosine: Chronic Pain

Neuropathic pain is a debilitating condition that affects millions of people worldwide, causing persistent pain in response to normally harmless stimuli. Despite its prevalence, effective treatments for neuropathic pain are limited due to a lack of understanding of its underlying chemical basis. Untargeted metabolomics (**Figure 9.8,** Nature Chemical Biology 2012) has

9. Metabolite Discovery

revealed that sphingomyelin metabolism was dysregulated in the spinal cord of rats suffering from neuropathic pain, specifically in the region associated with pain processing. Furthermore, it was discovered that an up-regulated metabolite called N,N-dimethylsphingosine (DMS) induced mechanical allodynia, a condition in which non-painful stimuli become painful. These findings provide insight into the chemical mechanisms underlying neuropathic pain and highlight the potential of targeting sphingomyelin metabolism as a therapeutic strategy.

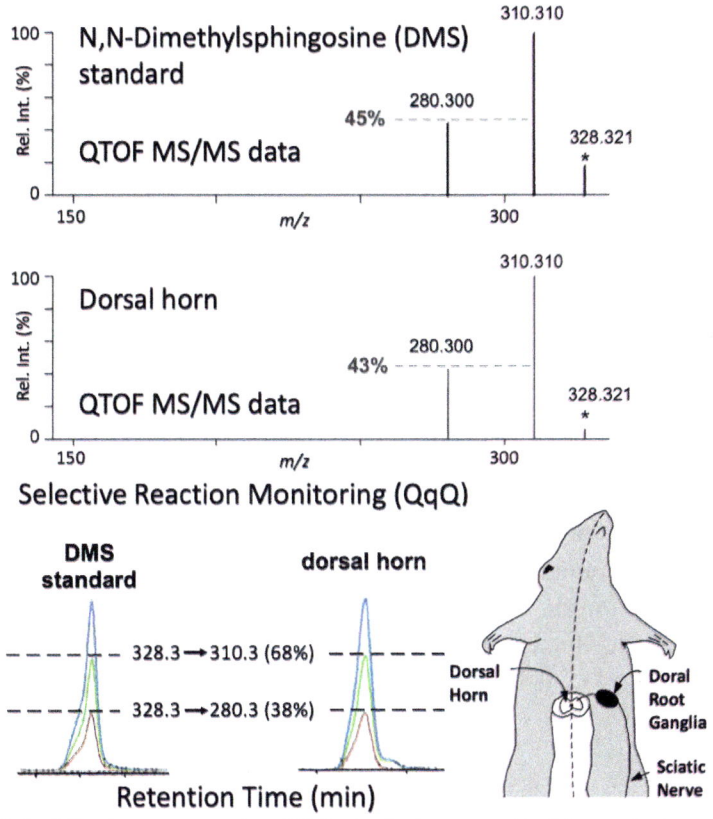

Figure 9.8. (top) Tandem mass spectra of DMS standard compared to data acquired from the dorsal horn of the spinal cord. This data was acquired from both QTOF and QqQ (bottom) instrumentation for additional validation.

Neuropathic pain is a challenging condition to treat, and current options often come with side effects and limited efficacy. While various molecular and cellular changes have been associated with neuropathic pain, the exact molecular causes remain unclear, hindering the development of effective treatments. To address this, the researchers employed mass spectrometry-based metabolomics to analyze tissue samples from rats with neuropathic pain. They focused on the chronic phase of pain and found that a significant number of metabolic changes occurred specifically in the spinal cord, suggesting its central role in maintaining pain sensitivity.

One of the most intriguing findings was the dysregulation of sphingomyelin-ceramide metabolism in the ipsilateral dorsal horn of the spinal cord, the region involved in pain processing. Sphingomyelin metabolism is known to play crucial roles in various cellular processes, including myelin formation and cell signaling. Dysregulated metabolites in this pathway, such as DMS, were found to be significantly up-regulated in the spinal cord of rats with neuropathic pain. Importantly, the researchers demonstrated that administration of DMS induced mechanical allodynia in healthy rats, confirming its role in pain development. Additionally, they observed that DMS triggered the release of inflammatory mediators from astrocytes, supporting the involvement of glial cells in neuropathic pain.

Overall, this study sheds light on the chemical basis of neuropathic pain and identifies the dysregulation of sphingomyelin metabolism, particularly the up-regulated metabolite DMS, as a potential contributor to pain development. The findings suggest that targeting DMS production or its downstream effects, such as inflammatory mediator release from astrocytes, could be a promising approach for the treatment of neuropathic pain. These insights open up new avenues for therapeutic intervention and provide hope for improving the lives of individuals suffering from this debilitating condition.

9. Metabolite Discovery

Identifying Metabolite N1,N12-Diacetylspermine:
Bacteria Biofilms Cancer

9. Metabolite Discovery

This was a relatively straightforward study, published in Cell Metabolism in 2015 (**Figure 9.9**), and was designed to investigate the role of colonic biofilms in colorectal cancer (CRC) development and their impact on the tumor microenvironment. Mass spectrometry-based metabolomics was used to analyze CRC tumors and normal-flanking tissues from 30 human patients, focusing on the identification of major metabolites employing untargeted metabolomics for discovery and targeted metabolite profiling for quantification validation.

The results showed that N1,N12-diacetylspermine were upregulated in all tumor tissues. However, its presence was significantly enhanced in patients with biofilm-positive CRC. Targeted metabolite profiling further confirmed its upregulation and other metabolites in the polyamine pathway. NIMS imaging provided spatial specificity of these metabolites in tumor versus normal tissues.

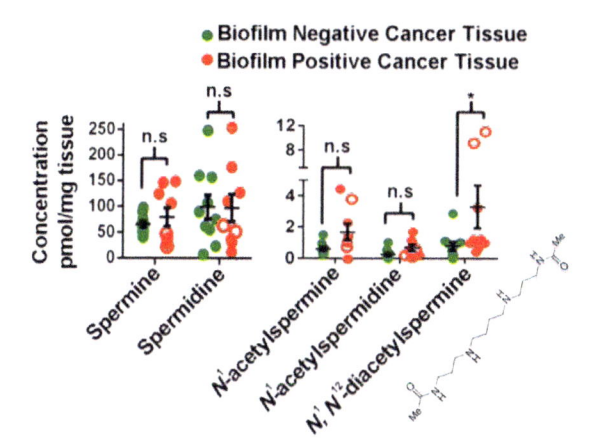

Figure 9.9. Targeted metabolomics concentrations of metabolites in cancers with (red) and without biofilm (green) (n.s.=not significant). Only N1,N12-diacetylspermine was deemed statistically significant.

The findings suggest that colonic biofilms can alter the tumor metabolome, resulting in increased levels of polyamines, which are known to regulate cell proliferation and tumor growth. The study highlights the potential biological significance of biofilms in CRC and their impact on the tumor microenvironment.

However, what I found especially intriguing about this study is that N1,N12-diacetylspermine is a well-known biomarker for at least ten cancers, this consistency implies a larger role for this molecule, possibly in immune suppression that the microbiome uses to stay viable.

Overview

Urea was the starting point in unraveling the complex puzzle of metabolism, and our understanding continues to grow. From simple molecules like urea to more intricate ones like acetyl-CoA, each discovery adds to our mechanistic knowledge. Metabolomics, the youngest of the omics fields, has played a vital role in driving these discoveries through advancements in analytics and computation. These technologies enable us to comprehensively characterize metabolites and their metabolism, leading to a deeper understanding of how they influence biological systems. The potential of these technologies extends beyond characterization; they hold the key to uncovering the biological activities of metabolites in health and disease. The concept of "Activity Metabolomics" (**Chapter 10**) aims to define the kinetics of metabolism and explore the biological effects of metabolites. As activity metabolomics emerges, it challenges the traditional central dogma of molecular biology by recognizing metabolites as powerful regulators of biological processes. These advancements pave the way for a new era of research, where metabolites take center stage as master manipulators of biology.

References

Giera, M., Yanes, O., Siuzdak, G., Metabolite discovery: Biochemistry's scientific driver. *Cell Metab. 34*, 21– 34 (2022) https://doi.org/10.1016/j.cmet.2021.11.005

Wikoff, W. R. et al. Metabolomics analysis reveals large effects of gut microflora on mammalian blood metabolites. *Proc. Natl Acad. Sci. USA* **106**, 3698–3703 (2009).

Yanes, O., Clark, J., Wong, D. *et al.* Metabolic oxidation regulates embryonic stem cell differentiation. *Nat Chem Biol* **6**, 411–417 (2010). https://doi.org/10.1038/nchembio.364

Patti, G., Yanes, O., Shriver, L. *et al.* Metabolomics implicates altered sphingolipids in chronic pain of neuropathic origin. *Nat Chem Biol* **8**, 232–234 (2012). https://doi.org/10.1038/nchembio.767

Lerner, R. A. *et al.* Cerebrodiene: a brain lipid isolated from sleep-deprived cats. *Proc Natl Acad Sci USA* **91**, 9505–9508 (1994).

Cravatt, B. F. *et al.* Chemical characterization of a family of brain lipids that induce sleep. *Science* **268**, 1506–1509 (1995).

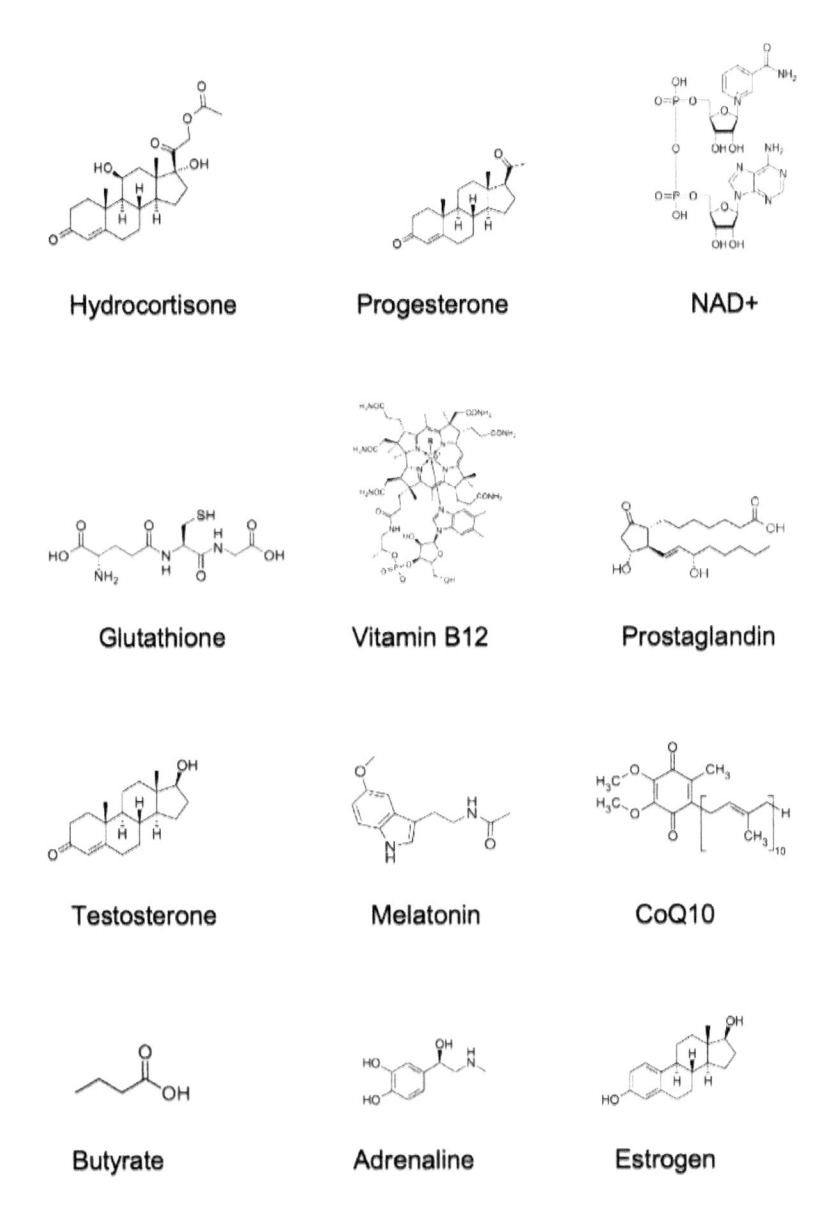

Hydrocortisone	Progesterone	NAD+
Glutathione	Vitamin B12	Prostaglandin
Testosterone	Melatonin	CoQ10
Butyrate	Adrenaline	Estrogen

Chapter 10
Metabolite Discovery

Perspective

Metabolite identification is a significant challenge and also presents a valuable opportunity in the field of biochemistry. The comprehensive characterization and quantification of metabolites in living organisms across the centuries (**Figure 10.1**) have laid the foundation for a vast biochemical knowledgebase, contributing to the resurgence of metabolism research in the 21st century. However, the identification of newly observed metabolites continues to be a persistent obstacle.

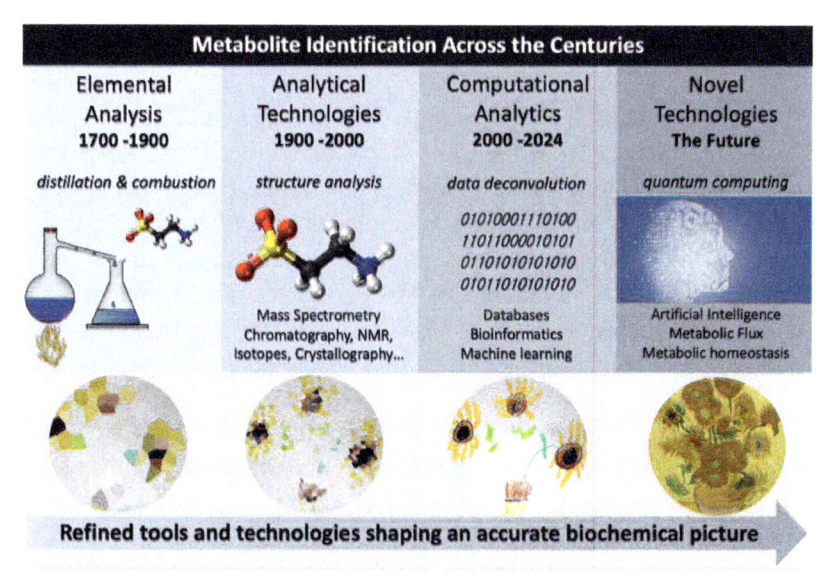

Figure 10.1. A broad depiction of the evolution of metabolic sciences and metabolite identification across the centuries as depicted by technology and (below) an art metaphor to van Gogh (credit Paul Giera).

10. Activity Metabolomics

In this final chapter, I describe an aspect of metabolomics that I have been engaged in now for three decades, namely identifying active metabolites (**Figures 10.1-10.3**). In particular we have been working toward identifying and understanding how to characterize which metabolites impact cellular physiology through modulation of other 'omic' levels, including the genome, epi-genome, transcriptome and proteome. This concept for identifying biologically active metabolites, or "activity metabolomics", and is already having broad impact on biology.

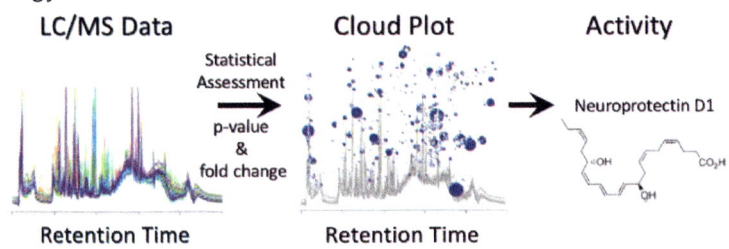

Figure 10.2. In its simplest form, going from dataset comparative analysis to statistical assessment, and finally to activity determination is the goal. Although identifying the best candidates and assessing (and understanding) activity can be time consuming.

The metabolome transcends the genome and proteome, representing the most downstream stage and the information flow through these different 'omic' levels of biochemical organization is described as the central dogma of molecular biology. Within this framework, the metabolome, has become widely accepted as the dynamic and sensitive measure of the phenotype at the molecular level, placing metabolomics at the forefront of biomarker and mechanistic discoveries related to pathophysiological processes.

However, the perception of metabolites mainly as a downstream product - biomarkers (of gene and protein activity) - has minimized the awareness of their far-reaching regulatory activity. Metabolite activity is a fascinating aspect of metabolism given that the metabolome interacts with and actively modulates all other 'omic' levels (**Figure 10.1**). Through this interaction metabolites also serve as direct modulators of biological processes and phenotypes. This concept has been investigated for decades, especially through the seminal discoveries of glucose, fatty acids and other lipids as regulators of insulin secretion and sensitivity and nutrient and energy sensing by the mTOR kinase. These findings have already shown the significant impact metabolites can have on biological systems.

Metabolite Profiling

Statistical Analysis — Metabolite ID

Identifying Metabolic Activity

Pathway Analysis — Network Analysis — Activity Databases

Metabolite Activity Screening

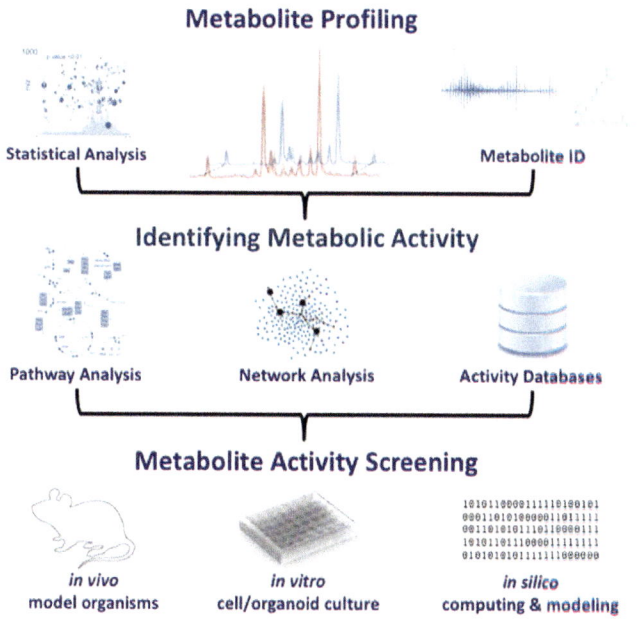

in vivo
model organisms

in vitro
cell/organoid culture

in silico
computing & modeling

Figure 10.3. (Top) Metabolomics-guided identification of bioactive metabolite candidate(s) starts with the comparative metabolome analyses based on the significance (p-value), amplitude (fold change) and direction of its change followed by identification. (Middle) Bioinformatics-based analyses with the use of metabolic pathway databases can be applied to assign the candidate metabolites to biochemical pathways. Protein and gene expression data can be further used to support and help in pathway annotation. (Bottom) Metabolites can be tested for bioactivity using a wide variety of assays (Tables 1 and 2).

How do "active metabolites" work?

The active metabolome drives phenotype modulation in a wide variety of ways. Oncometabolites are one of the best examples of active metabolites because of their early discovery and established mechanisms of phenotype modulation in cancer cells. The accumulation of these oncometabolites in distinct types of cancer cells is a causal process in malignant transformation. Oncometabolites, including D-2-hydroxyglutarate, L-2-hydroxyglutarate, succinate and fumarate, were found in cancerous tumors that had mutations in the enzymes (including Isocitrate

Dehydrogenase (IDH), Fumarase (FH) and Succinate Dehydrogenase (SDH) corresponding to the oncometabolites). The accumulation of the oncometabolites in the tumor cells arising from these mutations resulted in a proliferative cancer phenotype.

These oncometabolites are not only biomarkers of the diseases, depending on the activity of their respective enzymes, they can modify and interact with proteins and DNA thereby alter the proteome and the epigenome. Specifically, the activity of these oncometabolites stems from their inhibition of dixoxygenase enzymes resulting in a phenotype mirroring hypoxia, where the levels of transcription factor hypoxia inducible factor (HIF) are increased despite normal oxygen levels.

In addition to enzyme inhibition, other distinct biological activities for individual oncometabolites have been described. While the full scope of the biological effects of oncometabolites is still an active area of investigation, it has been demonstrated that they modulate protein-protein interaction, alter enzyme activity, lead to changes in the protein posttranslational modifications and modify the epigenome, all with the effect of propagating cancer. Recent mechanistic insights have revealed that metabolites strongly impact all layers of the omics landscape, from the genome, epigenome and transcriptome to the proteome. Within this framework the metabolome has two overarching mechanisms to control the functions of DNA, RNA and proteins: chemical modification and metabolite-macromolecule interaction that are further detailed in the following sections.

Metabolic modification of DNA, RNA and Proteins.

Metabolites drive pivotal covalent chemical modifications of proteins (post-translational modifications) and of DNA and RNA (such as methylation) which significantly affect cellular function (**Figure 10.4**). Post-translational modifications of proteins involve at least dozens of different small molecules that can be covalently bound to distinct amino acids as for instance lysine acetylation (derived from Acetyl-CoA) or cysteine palmitoylation (derived from Acyl-CoA). Many other metabolites are responsible for post-translational modifications including UDP-glucose for glycosylation, and itaconate for the alkylation of cysteine residues. Notably, abundance of many of these metabolite-induced protein modifications represent a powerful means of phenotype modulation.

DNA methylation (or a transfer of a methyl group from S-Adenosyl Methionine to cytosine) starts during embryogenesis and continues throughout our lifespan. Multiple metabolites, such as S-Adenosyl Methionine, glycine, pyruvate, galactose and threonine function as co-factors for posttranscriptional RNA modifications that act as sensors and transducers of information to control basic metabolic functions.

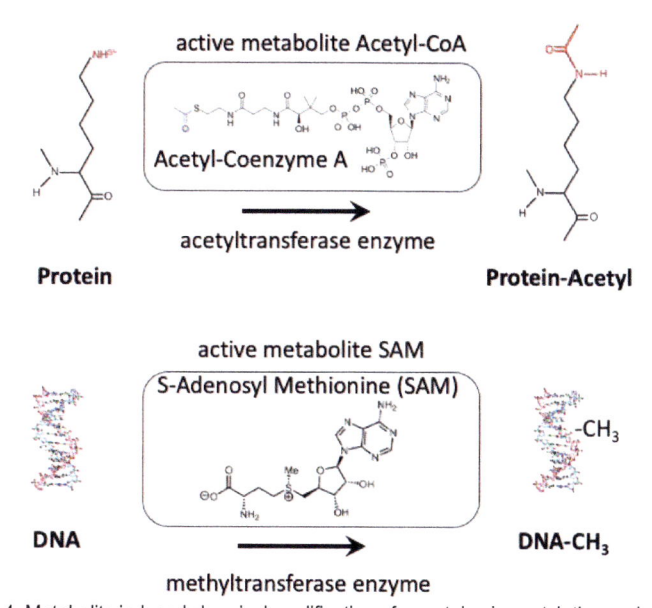

Figure 10.4. Metabolite induced chemical modification of a protein via acetylation and DNA via methylation.

Metabolite-macromolecule interactions

Cellular activity can be regulated through noncovalent interactions between metabolites and macromolecules. One way this happens is when a metabolite competitively binds to an enzyme's active site or at a different site, known as allostery (**Figure 10.5**). This concept applies not only to enzymes but also to various molecules like messenger RNAs, proteins, and G-protein coupled receptors (GPCRs), which are important for metabolite-activated signaling. GPCRs, for instance, play a crucial role in controlling blood pressure and other specialized cellular activations.

Transcription factors also participate in this regulation, determining how the system responds to cues by controlling the expression of specific genes. Riboswitches are involved in metabolite-controlled transcription and translation, with various metabolites, such as lysine and glutamine, acting as controllers.

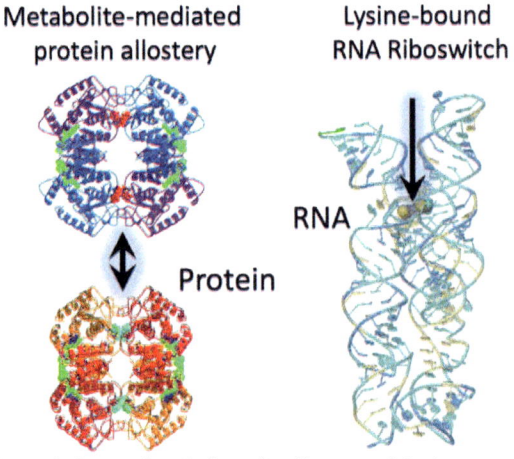

Figure 10.5. (left) Metabolite-mediated allosteric effect on a dehydrogenase protein and a (right) lysine mediate RNA riboswitch. Dehydrogenase adopted from Iwata et al. Nature Struct. Biol. 1994 and a RNA riboswitch .

Notably, glucose sensing in the brain plays a vital role in regulating hormone secretion, neuronal activity, and behavioral phenotypes related to feeding and energy expenditure. Interestingly, many active metabolites are typically considered common building blocks of cells, like amino acids for proteins, pyrimidine and purine bases for nucleic acids, and phospholipids for cell membranes.

Broader roles of metabolome in influencing global gene and protein activity have become more evident, for example dietary restriction in mouse models have shown profound changes in gene expression and improvement in aging phenotypes. It has also been found that the epigenome is influenced by metabolic status.

In another example, integrated analyses of yeast strains demonstrated that metabolite supplementation largely controls gene and protein expression, suggesting a system-wide regulation by metabolites.

Discovering active metabolites

Throughout history, metabolite detection, identification, and quantification have relied on biochemical approaches. But now, the field of metabolomics opens up new opportinuities for discovery at multiple levels. Unraveling the identity of active metabolites is a crucial aspect of this journey.

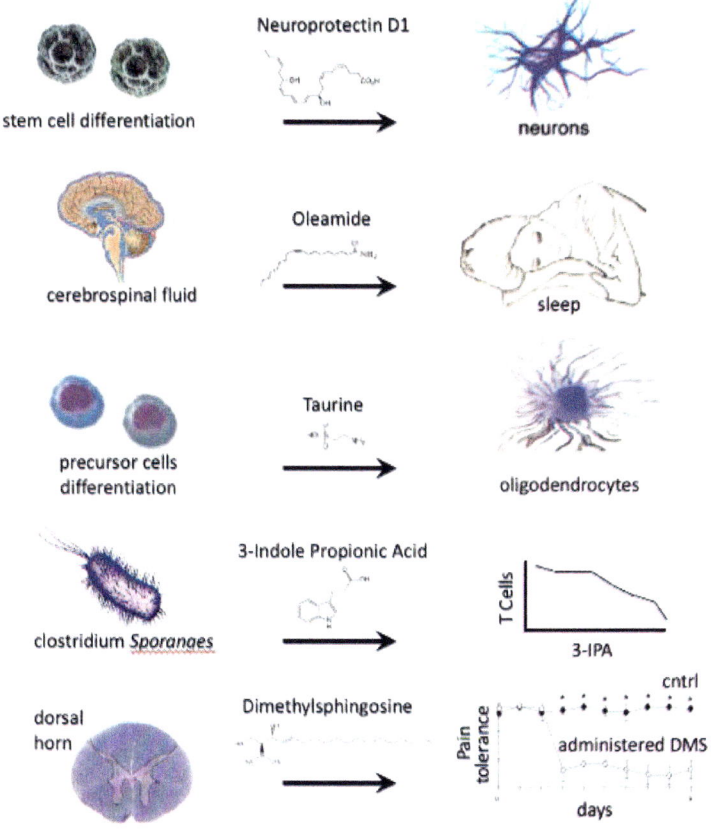

Figure 10.6. Metabolite activity demonstrated in stem-cell differentiation, oleamide observed in CSF during sleep deprivation, taurine enhances differentiation into mature oligodendrocytes, as part of a multiple sclerosis treatment, T-cell modulation by a metabolite derived solely from gut bacteria (3-IPA), and pain sensitivity amplified by DMS.

Mass spectrometry-based metabolomics, helps us identify features, annotate these features, statistically analyze them, and generate a list of potential metabolites. A number of peak detection and alignment softwares, like XCMS Online, MZmine2, Open-MS, and MS-DIAL, are currently available allowing for annotating features and identifying metabolites using databases METLIN.

Further to this, bioinformatic metabolic pathway and network analysis simplifies the complex data, helping us prioritize metabolites involved in distinct modules of the metabolic network can simply the candidate filtering process. And once the candidate lists are established, this metabolomics-guided activity screening is accomplished using in vivo and in vitro phenotypic, omics, and chemical biology strategies to identify the active metabolites and understand their mechanism of action.

Metabolic Activity Screening Strategies

Identifying active metabolites that modulate phenotype can be achieved through a variety of effective strategies. Utilizing metabolomics in conjunction with orthogonal molecular biology and computational approaches has proven to be successful in identifying active metabolites across various studies.

Table 10.1. General Principles for activity screening strategies

Technology	Mode	Principle
Cell-based assays	In vitro	Metabolite-induced cellular phenotype is quantified
Expression screening	In vitro	Metabolite-induced protein or transcript expression is quantified
Binding screens on protein arrays	In vitro	Metabolite-protein interactions are profiled
Chemical proteomics	In vitro	Metabolite-protein interactions are profiled
In vivo phenotype screening	In vivo	Metabolite-induced molecular and patho-physiological phenotypes of model organisms are quantified
Mathematical modeling	In silico	Mass balance or neuronal networks used to predict active metabolites
Cognitive computing	In silico	Active metabolites are predicted based on previous studies

In each of these examples, a crucial initial step involves implementing an appropriate screening strategy. These strategies encompass gene expression, protein expression, protein activity (such as enzyme activity), and the modulation of desired cellular phenotypes. For a comprehensive overview, please refer to **Table 1**, which outlines the principles underlying these assays.

In contrast to in vivo studies, cell-based assays offer several advantages, including high throughput capabilities and the ability to analyze various aspects of cell biology, such as cell morphology, biophysical function, and chemical properties. These assays also benefit from the use of molecular biology tools like fluorescent or luminescent reporters, providing valuable insights into cellular processes.

In the fields of metabolite-epigenome and metabolite-genome interactions, nucleotide sequencing approaches, such as chromatin immunoprecipitation and sequencing (ChIP-Seq), and bisulfite sequencing for detecting methylated DNA, are commonly employed. Additionally, chemical proteomics has emerged as a valuable tool to study metabolite-protein interactions. Techniques like protein arrays and thermal proteome profiling (TPP) allow researchers to examine the binding of specific metabolites to various proteins, offering a deeper understanding of cellular processes.

However, some of these screening approaches can be time-consuming, costly, and restricted to specialized laboratories. To efficiently identify potential candidates from metabolomics data, a combination of computational activity prediction strategies and chemical approaches can be used. This approach accelerates the assay process and enhances the selection of candidates for further investigation.

Advanced prediction methods, such as neural networks trained with large datasets from the ChEMBL database, are being adopted to predict the bioactivity of metabolites discovered through exploratory metabolomics. By leveraging this information, researchers can superimpose metabolomics data on activity constraints, optimizing the selection of metabolites for subsequent in vitro and in vivo bioactivity testing.

Additionally, knowledge from drug metabolism studies, which employ sophisticated in silico approaches to predict molecular sites prone to metabolic activation, can contribute valuable insights for these endeavors.

Overall, the integration of cell-based assays, nucleotide sequencing techniques, chemical proteomics, computational activity prediction, and drug metabolism knowledge offers a comprehensive approach to study metabolite interactions and identify potential bioactive candidates for further investigation. These combined efforts help streamline the screening process and facilitate the discovery of metabolites with significant biological relevance.

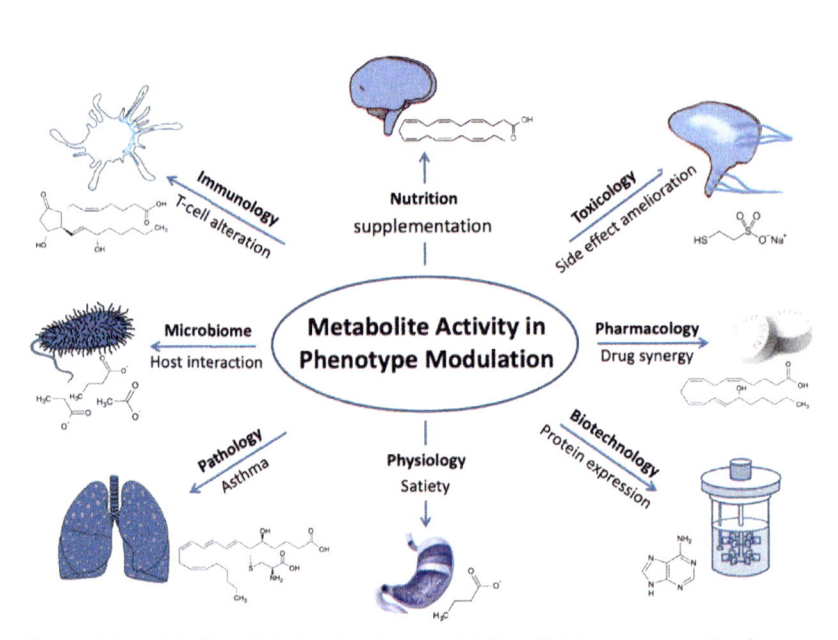

Figure 10.7. Metabolite activity for phenotype modulation. Metabolomics has already made a considerable impact in a wide variety of scientific areas through discovery of active, endogenous metabolites that can regulate different biological processes and thus modulate the phenotype in health and disease. This figure represents a variety of applications of endogenous metabolites ranging from nutrition to immunology.

Multi-omics integration for determining activity

In addition to metabolomics, the correlation and integration of multi-omic data present a useful approach to enhance the selection process of metabolites. By combining metabolite candidates with transcriptomic and proteomic data, researchers can gain deeper insights into selected pathways of interest. This multi-omic integration has been applied across diverse research areas, ranging from cancer metabolism to plant physiology and microbiology, yielding valuable information about distinct metabolite activities through quantitative modeling. The aim is to enable targeted interventions on specific pathways.

To handle the vast amount of data and reduce complexity, various mathematical approaches can be employed. For instance, metabolite and other omics data can be added to curated pathways, or novel pathways and fluxes can be modeled using the data. The field of multi-omics integration is continuously evolving, with multiple approaches being explored.

One common strategy in multi-omics integration involves using gene nomenclature linked to unique metabolite identifiers and combining existing pathway information (e.g., KEGG, Reactome, Biocyc pathways, Recon) with metabolomic set enrichments (available via platforms like MetaboAnalyst) and the newly developed ChemRICH platform. By utilizing this pathway information, researchers can effectively reduce complexity and filter out noise.

However, integrating the metabolome and other omic layers can pose challenges. Complex datasets are often acquired without a preset integration strategy. Addressing data robustness, sample harvest artifacts (batch effects), and distinct features of different omic datasets is crucial. Challenges in multi-omic data integration include noise removal, data prefiltering, matching various identifiers, selecting data dimensionality reduction methods, choosing computational approaches and mathematical models, model validation, and integration into trans-omics network modules. Efficiently advancing these approaches is an area of ongoing development.

Metabolic networks face validation challenges due to limited metabolite coverage resulting from analytical bias. Additionally, the full potential of omics data is not always utilized comprehensively, and data sharing strategies need improvement. However, multi-layered omic strategies, once acquired and integrated, can be highly useful, especially when applied to in vivo metabolism.

The future integration of genome-scale modeling, big data analysis, and machine learning strategies is anticipated to further prioritize metabolite activity in biological systems. In vivo and in vitro modeling studies have already demonstrated the power of this integrative omics approach, accurately predicting biological behavior based on multi-layered omic data acquisitions and metabolic phenotypes. Predictive pathway analysis within these platforms allows for straightforward and efficient metabolite mapping to background knowledge databases, be they curated reference pathway databases or genome-scale networks.

Table 10.2. Chemical biology and computational technologies.

Approach	Descriptions	Advantage	Limitations
Purification and isolation from a complex mixture	Fractionation of mixtures by chromatography and subsequent activity testing using assays	Universal, flexible biological assay	Tedious, signal overlap, minor components might be missed
Affinity selection mass spectrometry	Incubate metabolite mixture and target enzymes/proteins, size exclusion of bound components, MS based ID	Universal approach, no protein immobilization necessary	Non-specific binding may be obtained. Ligand binding and not activity is assessed.
Affinity purification LC/MS	Affinity based protein purification & MS based ID of components from pulldown of complex metabolite mixtures	Universal approach which can be used in vivo (e.g. yeast cells)	Antibody-dependent. Ligand binding and not activity is assessed.
Thermal proteome profiling	Binding of a ligand to a protein in vivo or in vitro results in increased thermal stability	Universal approach, physical stabilization of proteins	Low thru-put, non-specific binding & binding is not bioactivity
Metabolite	Integration of	ID of	Long-term,

profiling combined with molecular biology	metabolomics with orthogonal molecular biology experiments	mechanism of action of a signaling metabolite	fastidious
Integrated network analysis	Combination of transcriptional and metabolomics data for the identification of active metabolic sub-networks	Network analysis allowing for a systems wide comparison	Mandatory, preassembled metabolic networks (species dependent)
Flux balance analysis	Mathematical in silico approach for the calculation of the metabolic flux through a network.	Easily computable. No kinetic parameters needed.	In silico genome metabolic network predict steady state
Metabolite set enrichment and network analysis	Computational based on overrepresentation & probability analysis of metabolomics to ID active pathways	Rapid, allows for direct association with relevant information	In silico approach, significant amount of false positive metabolite IDs
Bioactive Natural Products using Networks	Molecular networks embedding known bioactivity & other data to highlight potentially bioactivity	IDs bioactive compounds using fragment similarity	Structural elucidation of compounds remains a challenge

It is worth noting that current metabolic spectral databases only cover a fraction (up to 60%) of genome-scale metabolism, indicating a limitation due to the "dark" metabolome.

Ultimately, these integrated approaches aim to generate accurate biological models, allowing the identification of the most promising candidates for activity screening. The multidimensional integrated "omic" landscape offers exciting opportunities and unconventional solutions for predicting metabolites that modulate phenotype.

Perspective: Applications of activity metabolomics

Activity metabolomics has a broad range of applications, impacting phenotypes across various organisms, from simple prokaryotes to complex human physiology. Its versatility is exemplified by its ability to enhance biotechnological applications, such as boosting Bordetella pertussis vaccine and protein production in E. coli. Additionally, activity metabolomics plays a significant role in modulating the microbiome, where alterations to the mammalian metabolome can influence microbial communities, with microbiome-derived metabolites impacting immune cells and satiety.

In the emerging field of immunometabolism, specific metabolites like prostaglandin E2 have been found to influence immune responses, promoting certain T cell types while suppressing macrophage and neutrophil activity. At an organ level, metabolites like leukotrienes control pathophysiological reactions, such as in asthma. In complex organisms, nutritional interventions, like omega-3 fatty acid supplementation, have shown multiple benefits without severe side effects.

In pharmacology and toxicology, metabolites are administered to reduce toxicity, such as using ethanol to counteract methanol poisoning or administering scavenger metabolites to mitigate the effects of certain chemotherapeutics. Drug synergies with endogenous metabolite classes are also utilized for therapeutic purposes. Even simple metabolites like glutamine have proven effective in treating complex diseases like sickle cell disease, as evidenced in phase 3 clinical trials.

These examples highlight the wide applicability of active metabolites in modulating biological processes, cellular metabolic states, cell activation, differentiation, proliferation, and complex tissue functions. Notably, the effect of a metabolite is context-dependent, and its induced phenotype can vary significantly based on the biological system it is applied to. The goal of activity metabolomics is to provide a framework to understand and systematically quantify these phenotypic changes.

One such example is α-ketoglutarate (AKG), an active metabolite that alters phenotypes in a context-dependent manner. It regulates glucose metabolism and uptake in bacteria, extends the lifespan in Caenorhabditis elegans through mTOR inhibition, supports regulatory T-cell differentiation from Th1 cells in immune cells, and enhances tissue and muscle regeneration in humans via processes involving ERK and other factors. Its binding to the G-protein-coupled receptor Oxgr1 translates into increased transporter synthesis and hypertension through salt reabsorption in the kidney. An overarching theme in these settings may be the modulation of anabolic cell activity, which could be considered the "common denominator" of AKG's role in the activity metabolome.

The current challenge in activity metabolomics is to link metabolites systematically and quantitatively to an organism's phenotype ("phenome"). Achieving this goal will necessitate comprehensive metabolomics and phenotypic data, as well as the computational integration of other omic data. Through such efforts, activity metabolomics holds promise in unraveling the intricate connections between metabolites and phenotypes, contributing to a deeper understanding of biological systems, and paving the way for innovative therapeutic strategies.

Conclusions

Increasing attention is being drawn to the idea that metabolites can act as controllers, not just cogs in a system. Termed "activity metabolomics," this approach employs metabolomics to identify active metabolites. The central focus is to identify master metabolites, achieved through computational integration of metabolomics, systems biology, and bioactivity data. This enables the identification of potent metabolites that modulate biological processes and cell physiology.

Many challenges, such as metabolite identification and annotation, have already seen significant progress. **Figure 10.1** illustrates the goal: leveraging metabolomics-driven screening methods to identify these master regulators. Once accomplished, activity metabolomics stands to impact multiple scientific disciplines.

10. Activity Metabolomics

Endogenous Metabolite Success Stories

On a final note, it is worth mentioning a few of the many success stories for highly recognizable endogenous metabolites. These clinically relevant endogenous molecules each generate hundreds of millions to over ten billion dollars a year in revenue.

Hydrocortisone

Progesterone

NAD+

Glutathione

Vitamin B12

Prostaglandin

Testosterone

Melatonin

CoQ10

Butyrate

Adrenaline

Estrogen

References

Rinschen, M.M., Ivanisevic, J., Giera, M. *et al.* Identification of bioactive metabolites using activity metabolomics. *Nat Rev Mol Cell Biol* **20**, 353–367 (2019). https://doi.org/10.1038/s41580-019-0108-4

Guijas, C., Montenegro-Burke, J., Warth, B. *et al.* Metabolomics activity screening for identifying metabolites that modulate phenotype. *Nat Biotechnol* **36**, 316–320 (2018). https://doi.org/10.1038/nbt.4101

Giera, M., Yanes, O., Siuzdak, G., Metabolite discovery: Biochemistry's scientific driver. *Cell Metab.* *34*, 21–34 (2022) https://doi.org/10.1016/j.cmet.2021.11.005

Index

A
a.j. dempster 3
accuracy 62, 99
acetonitrile/h2o 40
alan g. marshall. 4
apci 43
appi 45
aprotic cosolvents 40
atmospheric pressure 43
atmospheric pressure chemical
ionization 43
atmospheric pressure
average mass 63, 95, 96

B
barber, michael 34
biomolecular mass spectrometry
.... 34

C
calculating molecular weight average
mass. 95
calibration 95, 99
calibration compounds 101
capillary columns 28
carbohydrates. 25, 36, 45, 61, 102
.... 107
cationization 31
charge (or inductive) detector 85
chemical ionization (ci) 54
chlorine 97
cleaning 102
collision gas 67, 78, 134
collision-induced dissociation (cid)
.... 11, 67, 122, 134
cooks, graham 34
coulombic repulsion 36

D
de novo sequencing 69
declustering 127
delayed extraction (de) 73
deprotonation 30
detector 83
direct infusion 28
direct insertion 28
direct insertion probe 52

double focusing magnetic sector
.... 70
Double sector 79

E
electron capture 32, 33, 54, 56
electron capture ionization 32
electron ejection 32
electron ionization 34, 52
electron multiplier 83
electrospray 11, 108
electrospray ionization ... 3, 5, 153
electrospray solvents 40
endogenous metabolite 240
exact mass. 96

F
fab matrix 52
faraday cup 60
fast atom bombardment 51
fenn, john b 5
fourier transform icr mass
spectrometry 4
fragmentation 5, 167, 182, 237
francis w. aston 2
franz hillenkamp 6, 34

G
gas chromatography electron
ionization mass spectrometry. 71

gas chromatography mass
spectrometry 29, 53
gel electrophoresis 129, 137

H
helium 64, 67
henry benner 8
high performance liquid
chromatograph 130
high resolving power ... 8, 75, 78,
.... 96
historical developments 18, 19

I
Ion cyclotron resonance m 4, 75
ion detector 26, 27, 83, 28
ion source 34
ion trap 67, 69
ionization ... 5, 25, 29, 33, 34, 41,
.... 43, 45, 52, 54, 57, 107, 108

ionization mechanism 108
ionization method 29
ionization sources 34, 56

J
john a. hipple 4

K
karas, michael 6, 34, 45
kinetic energy (ke). 54, 70

L
lc ms/ms 143
lc/ms 126
lc-maldi ms/ms 148
linear ion trap 69
linear time-of-flight (tof) 71
liquid chromatography mass
spectrometry 178

M
maldi matrix 46, 58, 108
maldi tof 71, 72, 136
maldi-ms ... 6, 27, 45, 129, 142
maldi-tof reflectron ... 115, 121
mapping 129, 146, 149, 151
mass analysis 61
mass analyzers 60
mass range 63
mass spectrometer 26
mass spectrometry 1
mass-to-charge ratio 26, 37, 47
matrix-assisted laser
desorption/ionization (maldi). 5
melvin b. comisarow 4
metabolite   8, 9, 13, 91, 167, 169
metlin 192, 193, 196, 198, 201, 203, 205, 207
monoisotopic mass. 63, 95, 96
multiple charging 38, 123

N
nanoesi 41
nominal mass 96

O
on-plate sample wash 118
orifice-induced 127

P
paul, wolfgang 4, 67
peptide mass mapping 129, 130
peptides and proteins 113

photomultiplier conversion
dynode 84
post translational modifications
.... 114, 131, 133, 165, 228
protein databases 133
protein digests 117
protein id 142
protein mass mapping ..129, ... 146, 149
protein profiling 141
proteins 115
protonation 30, 33, 108

Q
quadrupole 65
quadrupole ion trap 67
quadrupole time-of-flight. 74
quantitation 92, 93

R
resolution 62, 82

S
sample inlet. 27, 58
sample preparation. 95, 108
.... 117, 166
sample separation 136
scan speed 64
sds-page 137, 138, 140
soft ionization –
electrospray ionization ms.
.... 5
solubility 98
Surface-based ionization
techniques 115

T
tanaka, koichi 5, 34
tandem mass analysis 64
tandem mass
spectrometry.71, 73 time-
of-flight (tof) mass
analyzers 47,71

V
vacuum 87

X
xcms 173, 175, 177, 178, 179,180, 181, 183, 184, 185, 187, 188, 189, 190

Z
ziptip™ sample wash 118

Things to do during a pandemic: cut hair, ride bikes, build stuff, write a book.

Cite:
Siuzdak, G.,
Activity Metabolomics and Mass Spectrometry;
MCC Press: San Diego, USA, 2024.
https://doi.org/10.63025/LCUW3037

Printed in Great Britain
by Amazon